"This is one of the most brilliant books I have ever come across. Its aim is to address the question of what is doubt, not directly in any circumscribed definitional standard, but by examining the powerful role it has had (and continues to have) in human life. To do so, Geoff Beattie takes the reader on a narrative journey into the minds of some of the greatest writers and scientists of history in order to cast light on how their doubts shaped their creative and scientific accomplishments. He also looks at how doubt is harnessed maliciously to spread conspiracy theories about such things as climate change. Written in his usual accessible narrative style – let us not forget that Beattie is also an accomplished novelist as well as a distinguished psychologist – this book will be hard to put down. Its implications for grasping who we are as a species – we are the only species that possesses doubt – and for harnessing or counteracting its enormous power of control over human life, make it a *must* read for everyone."

Marcel Danesi, PhD, FRCS, *Professor Emeritus of Anthropology, University of Toronto, Canada*

"Beattie brilliantly illustrates the science of doubt with fascinating case studies from doubters like Kafka, to non-doubters like Picasso and how it can be addressed therapeutically, as in Brendan Ingle's boxing gym in Sheffield."

Brian Butterworth, *Professor Emeritus of Cognitive Neuropsychology, University College London, UK*

"Geoff Beattie has written a brilliantly entertaining book about the little considered phenomenon of doubt, focusing mainly but not exclusively on self-doubt. Part memoir, part an examination of the psychology of doubt, and part an examination of the role of doubt (or lack of it) in the lives of key historical figures – Jung, Kafka, Picasso, Ernest Dichter (a psychoanalyst who devoted his talents to promoting the consumption of cigarettes) and Alan Turing – the book offers a wide-ranging and unique appreciation of the importance of doubt in individual minds and in human affairs."

Richard Bentall, PhD, FBA, *Professor of Clinical Psychology, University of Sheffield, UK*

Doubt

Blending the latest academic research with case studies of famous figures, this highly insightful book presents 'doubt' as a central concept for psychology. It is a concept which has been oddly neglected in the past, despite its ubiquitous nature and far-reaching influence.

Exploring everything from self-doubt and impostor syndrome to the weaponisation of doubt with respect to climate change and the marketing of cigarettes, bestselling author Geoffrey Beattie navigates readers through the various ways doubt can start and develop, changing the individual in the process. Written in Beattie's distinctive and engaging style, *Doubt* takes the reader into the lives of transformational thinkers, artists, scientists and writers to explore how and why doubt was crucial in their lives and how the likes of Kafka, Jung, Picasso and Turing succumbed to doubt or learned to control it. Beattie argues that doubt is central to the self; it can be either a safeguarding mechanism or a distraction, rational or irrational, systematic or random, healthy or pathological, productive or non-productive. The book helps readers to recognise how doubt may have been operating in their own lives and to identify how and when it has been used against us – for example, to prevent climate action – and at what personal and societal cost.

Presenting a compelling case for why doubt cannot be ignored, this book is of major interest to academics from a wide range of disciplines, including social and cognitive psychology, clinical and counselling psychology, sport psychology, sociology, business studies, politics, art and literature, as well as the general public, who may well see something of themselves in its pages.

Geoffrey Beattie is a prize-winning academic, author and broadcaster. He received his PhD from Trinity College Cambridge and is now Professor of Psychology at Edge Hill University, as well as Fellow of the British Psychological Society, the Royal Society of Medicine and the Royal Society of Arts.

Doubt

A Psychological Exploration

Geoffrey Beattie

LONDON AND NEW YORK

Designed cover image: Getty Images

First published 2023
by Routledge
4 Park Square, Milton Park, Abingdon, Oxon OX14 4RN

and by Routledge
605 Third Avenue, New York, NY 10158

Routledge is an imprint of the Taylor & Francis Group, an informa business

British Library Cataloguing-in-Publication Data
A catalogue record for this book is available from the British Library

Library of Congress Cataloging-in-Publication Data
Names: Beattie, Geoffrey, author.
Title: Doubt : a psychological exploration / Geoffrey Beattie.
Description: 1 Edition. | New York, NY : Routledge, 2023. | Includes
    bibliographical references and index.
Identifiers: LCCN 2022026173 (print) | LCCN 2022026174 (ebook) |
    ISBN 9781032252056 (hardcover) | ISBN 9781032252049 (paperback) |
    ISBN 9781003282051 (ebook)
Subjects: LCSH: Belief and doubt. | Self-confidence.
Classification: LCC BF773 .B423 2023 (print) | LCC BF773 (ebook) |
    DDC 153.4—dc23/eng/20220825
LC record available at https://lccn.loc.gov/2022026173
LC ebook record available at https://lccn.loc.gov/2022026174

ISBN: 978-1-032-25205-6 (hbk)
ISBN: 978-1-032-25204-9 (pbk)
ISBN: 978-1-003-28205-1 (ebk)

DOI: 10.4324/9781003282051

Typeset in Bembo
by Apex CoVantage, LLC

For Joshua and Olivia Beattie and a future full of hope and promise, and less doubt.

Contents

Acknowledgements   x

1 Introduction: the nature and meaning of doubt   1

2 Alone in doubt   17

3 Jung's dream   29

4 Feeling like a fraud   40

5 'I, the King'   57

6 The perils of a doubt-free life   75

7 Treating doubt the hard way   86

8 Manufacturing doubt   106

9 Our house is on fire   136

10 Concluding remarks   149

References   154
Index   160

Acknowledgements

I would like to thank Edge Hill University for proving such a pleasant and rewarding environment in which to work, especially John Cater, George Talbot and Rod Nicolson. The *Times/Sunday Times* 'Modern University of the Year' 2022 is the most recent accolade with so much more to come. This is a university with real potential. The psychology department has a wonderfully supportive academic culture, I have many great colleagues in the department (including my closest colleague Laura McGuire), and I feel very fortunate to be working there. My agent Robert Kirby from United Agents and I discussed the concept of doubt before I began (he has a great sense of what should be pursued), and I must thank him for his support over many years, as well as Eleanor Taylor, my editor at Routledge, who is a pleasure to work with on my books. I have written about smoking before in the chapter on conflicted habits in *The Conflicted Mind*, also published by Routledge, and I thank them for permission to use this material here.

1   Introduction

The nature and meaning of doubt

Doubt is a lack of confidence or uncertainty about something or someone, including the self. It is central to science, the law, ethics, politics and philosophy which all involve elaborate and often refined adversarial processes to promote, consider and evaluate doubt in the light of the available evidence. Descartes used Cartesian doubt, the process of being sceptical or doubting the truth of one's beliefs, as a methodological tool in his philosophical investigations. But it is also central to the self. It can be a safeguarding mechanism or a distraction, it can be rational or irrational, systematic or random, healthy or pathological. Obsessive-compulsive disorder is sometimes referred to as a 'disease of doubt'. According to Freud, obsessional neurotics have a 'need for uncertainty in their life, or for doubt' (Freud 1909/1996: 232). They are paralysed by ambivalence, immobilised by two instinctual impulses, love and hate, directed at the same object.

In a famous letter to Lou Andres-Salome, Freud speculated on the aetiology of doubt as a symptom in his famous case study of the 'Rat Man'. This nickname, given to the lawyer Ernst Lanzer by Freud, related to Lanzer's recurrent, involuntary and nightmarish thoughts about rats which began after he heard an account told to him by a fellow army officer, concerning a particularly gruesome torture method said to be used by the Chinese on prisoners. It involved a rat being trapped in a pot and fastened to the rear of the prisoner; the rat would then gnaw through the captive's anus to survive. This was the indelible image that Lanzer had in his head, and he developed obsessional thoughts about a similar fate befalling both his wife and his father. Freud's hypothesis was that these obsessional and gruesome thoughts resulted from a conflict between loving and aggressive impulses towards both individuals. The obsessional symptoms meant that the patient avoided resolving this conflict, subsequently failing to resolve other difficult decisions between alternative positions and courses of action throughout his life. It thus became a life marked by doubt and indecision. In terms of its psychological origin, Freud hypothesised (in somewhat typical fashion, it must be said) that the symptoms derived from the patient's earliest sexual experiences, in particular the harsh punishment he received by his father for childhood masturbation (but see Mahony 1986). 'The tendency to doubt . . . is the continuation of the powerful ambivalent tendencies in the pregenital

DOI: 10.1324/9781003282051-1

phase, which from then on become attached to every pair of opposites that present themselves' (Freud and Andreas-Salome 1966/1972: 77). This is one position on the nature and origin of doubt but not necessarily the one with the strongest evidence.

Doubt is an instrument of rational thought – in science, law, philosophy, everyday thinking on the one hand and on the other some sort of psychological problem, perhaps even a symptom of major psychological dysfunction rooted in the earliest stages of psychosexual development (at least according to some). Doubt is clearly a very broad concept. Doubt is also a great driver, perhaps the greatest driver of all, *and* a great inhibitor when it comes to human action. It drives science, judicial decisions, philosophical understanding, progress, change, positive action, but it can also inhibit decision-making and stop change and lead to procrastination, worry, superstition and delay. It is internal and conscious ('a perception of indecision') and therefore highly personal. It is often closed off, something we often choose not to share with others, even those closest to us. Imagine how your partner would feel if they knew about your doubts about the relationship? Or vice versa – if you knew about their doubts?

Over the years, writers, novelists, biographers and historians seem to have had as much or more to say about doubt than psychologists (with perhaps one or two notable exceptions). It is a core part of our mental lives, and we enjoy getting a glimpse of its structures and functions in literature, to understand its complexity and deviousness (at times). Freud, as we have noted, viewed doubt as a 'symptom', a mark of resistance, an indication to the analyst of the significance of the repressed element to which it is related. Jung, also from a psychoanalytic perspective, viewed it very differently. He wrote in his letters, 'Doubt and insecurity are indispensable components of a complete life' (Jung 1951/1976). In this book we will consider amongst other things how doubt fitted into Jung's own life and why he viewed it in the way he did, but, of course, many of the more 'neurotic' aspects of doubt will also be considered in detail.

Many other psychologists ignore doubt completely; they talk more frequently about perception of risk (which is not the same), uncertainty, anxiety, worry, apprehension, self-efficacy, impostor syndrome and fear of failure, but they do not link these up to talk about doubt itself – that core part of our mental life. This is a shame. It is a major part of our everyday social experience, a major part of who we are. And doubt is everywhere. From that nagging unpleasant thought at the back of our minds when we're having to choose a pair of socks (but that might just be me) to that 'Eureka!' moment based on a process of more prolonged and productive doubt when a solution suddenly pops into our head. Without that process of conscious and reflective doubting the solution might never have arisen. From the child crying for its mother, seeming to 'doubt' whether she'll ever return (at least inferred on the basis of the surprise when she sees her mother again, although admittedly it is very odd to talk about the experience of doubt in pre-linguistic infants), to Jesus himself, abandoned on the cross, his anguish bathed in doubt (although I'm starting to doubt whether that's a good example because it might offend some Christians). But that is the breadth of the subject.

Doubt can seemingly start in one place in time, in one modality, and spread across time and across the whole functional architecture of the brain, from a slight feeling of unease to major discomfort, and then to active thinking about the potential source or target of the doubt, gradually becoming part of how we deal with the world, and thereby part of our personal and social identity. Doubt doesn't have to work in this linear way with emotion arising first, leading ultimately to cognition and conscious awareness. With doubt, cognition and emotion are intimately connected, one or other can take precedence. We can also often notice our doubting, and this may be important to how we view ourselves. Many from both inside and outside psychology suggest that the roots of doubt can usually be traced to our earliest socialisation experiences, if not necessarily critical moments in our psychosexual development (as Freud hypothesised), particularly our interactions and relationships with parents. Others suggest that doubt can arise from a range of important interactional and emotional experiences throughout the course of one's life. There are great thinkers, artists and writers on both sides of the argument, as we will see, so it's probably best to start with an open mind.

We know that Kafka suffered from extreme self-doubt and clearly believed that he understood where all this doubt and self-doubt originated from. The author of such classics of literature as *The Metamorphosis, The Trial* and *The Castle* and regarded by writers like Auden, Nabokov and García Márquez as one of the principal figures of twentieth-century literature, was crippled with doubt throughout his life. He wrote a letter to his father in November 1919, explaining why even as a child on passing his examinations and even being awarded prizes for his work, he had always felt like 'an embezzling bank clerk who, still at his job and trembling at the thought of being discovered, takes interest in the petty routine business of the bank that he still has to see to' (Kafka 1919/2011: 55). This is what we would now call 'impostor syndrome'. He describes how he was drowning in self-doubt, tormented by his thoughts, fantasies and dreams. Fantasies since he was a child that after he had passed his school examinations that he would inevitably be caught out – discovered by his teachers to be 'the most incapable and . . . the most ignorant of all' so that they would 'instantly spew me out, to the jubilation of all the righteous, now liberated from this nightmare' (Kafka 1919/2011: 54). Fantasies and thoughts that never left him.

Kafka makes the origin of this self-doubt clear in his great accusatory letter – it derives, he writes, from the emotional abuse that he, a 'timid child', suffered at the hands of his overbearing father, Hermann, a self-made man with an 'unlimited confidence in your opinion' in Kafka's accusing words. 'It is only as a father that you were too strong for me, especially since my brothers died young, my sisters only arrived much later, so that I had to endure the first knock or two all alone, and for that I was much too weak' (Kafka 1919/2011: 9–10). He cites his father's chronic disapproval, his 'belittling judgments', his indifference to his son's intellectual and creative interest, his threats, his dismissive attitude to his son's work, the absence of love as the sources of his own doubting, which left him unable even to think properly – 'it was almost impossible to endure this and still work out a thought with any measure of completeness and permanence.'

He was intimidated by his father both physically and mentally:

> I remember, for example, how we often undressed together in the same changing room. I was skinny, weakly, slight; you were strong, tall, broad. Even in the changing room I felt pitiful, and what's more, not only in your eyes, but in the eyes of the entire world, for you were for me the standard by which everything was measured.
>
> (Kafka 1919/2011: 15)

In terms of the father's behaviour, Kafka writes that 'Your extremely effective rhetorical child-rearing devices, which never failed with me, at least were abuse, threats, sarcasm, spiteful laughter and strangely enough – self-pity' (Kafka 1919/2011: 23). You can also see the prototype of the Kafkaesque situation in what he said he had to endure:

> I was forever in disgrace: either I obeyed your orders, that was a disgrace, for they were, after all, meant only for me; or I was defiant, that was also a disgrace, for how dare I defy you; or I could not obey because I didn't have, for example, your strength, your appetite, your skill, even though you expected it of me as a matter of course; and that was the greatest disgrace of all.
>
> (Kafka 1919/2011: 20)

This is clearly a double bind (Bateson 1973) – the impossible situation where any response gets punished. Not a double bind in the 'classic' sense of an utterance with two simultaneous (and objectively observable) contradictory channels of communication, usually involving the verbal and nonverbal channels where any possible response gets punished. In Bateson's words, 'A secondary injunction [often communication through nonverbal means like posture, gesture or tone of voice] conflicting with the first at a more abstract level, and like the first enforced by punishment or signals which threaten survival' (Bateson 2000: 207). Rather it is a double bind realised over time involving the interpretation of multiple communications (sometimes just involving contradictory verbal statements alone) resulting in the feeling that no rational response is possible (Beattie 2021) and that any response inevitably leads to punishment – disgrace or worse in this case. Kafka hesitated in this relationship with his father, both literally and metaphorically:

> I acquired in your presence – you are, as soon as it concerns your own matters, an excellent speaker – a hesitant, stammering manner of speaking, and even that was too much for you, eventually I kept silent, at first perhaps out of spite, and later because I could neither think nor speak in your presence.
>
> (Kafka 1919/2011: 22)

The doubt spread to every aspect of his existence. 'I lost confidence in my own abilities. I was unsteady, doubtful. The older I got, the more material you could hold against me as proof of my worthlessness; gradually, in a certain regard, you began actually to be right' (Kafka 1919/2011: 23). He even had doubts about his own body, and this seriously affected his health:

> But since there was nothing at all I was certain of, since I needed a new confirmation of my existence at every instant, since nothing was in my very own indubitable sole possession, determined unambiguously only by me, in truth a disinherited son, naturally I became unsure of even the thing nearest to me, my own body.
>
> (Kafka 1919/2011: 23)

This was the foundation for 'every form of hypochondria' that Kafka developed, a range of anxieties about his digestion, about his hair falling out, about his spinal curvature and on and on, until he finally succumbed to a genuine illness, the tuberculosis that killed him at forty. He died of starvation in the end as the condition of his throat, tightened and closed by tuberculosis, made eating too painful. This has not stopped some psychologists from suggesting that the starvation was a form of anorexia, forgetting for a moment the pain of swallowing that he was experiencing at this time.

Kafka was thirty-six years old when he wrote the famous letter. He gave the letter to his mother Julie to pass on to his father, but she returned the letter to her son unread by the intended recipient, perhaps feeling that the letter would not accomplish what her son hoped for, and that he would be crushed again. It had taken Kafka thirty-six years to explain to his father how his father's behaviour made him feel and to confess his inhibited and secret self-doubts, the doubts that only leaked out through his avoidance behaviours towards his father, his neuroses and his hypochondria. But, of course, these doubts underpinned his art (which his father had always summarily dismissed). The writing was his 'attempts at independence, attempts at escape, with the very smallest of success' (Kafka achieved little success in his own lifetime). The doubts, articulated and explained, hold the key to understanding what it is liked to be trapped in a complex and contradictory 'Kafkaesque' situation. By giving us a word for it, unintentionally of course, he changed our views of the absurd, complex, bizarre and illogical; he allowed us to name it and thereby recognise and report it.

His father never read or heard the confession. I've always thought that was a real pity. Hearing about doubt and understanding it may help us understand each other, it may help us connect and feel less alone, even in disintegrating or dysfunctional relationships like this one.

There has been such emphasis in the past few decades on the articulation and sharing of emotional experiences and their importance for our mental and physical well-being (Pennebaker 1989, 1993, 1995, 1997, 2000, Lee and Beattie 1998, 2000). But we can express emotions whilst keeping our doubts covered and closer to our chest. Doubts reverberate inside, they resonate and grow.

We feel like frauds, impostors, cheats, and we never tell anyone. We doubt the efficacy of the Covid-19 vaccine, and we're frightened to admit it – we just stay quiet and don't get vaccinated. Easier that way. I stop in the changing room of my gym, and turn around four times to check my locker, doubting that I had picked all my belongings up. Somebody notices this odd action. I see him looking. 'I've left things behind before,' I say. This is not true, but I must explain myself. I always check like this. My own personal doubts have become public and need to be explained, they have become obsessive, odd.

The American Psychological Association defines doubt as 'a lack of confidence or an uncertainty about something or someone, including the self' – they remind us that doubt is a 'perception' but typically 'with a strong affective component'. I know that feeling intimately, many of us do, but perhaps not *everybody*. Some seem immune to doubt. I have experienced that affect many times. It's an unsettling sort of feeling that comes from nowhere; it can be very intense. I hesitate in choosing between alternatives, endlessly weighing up the pros and the cons, until eventually arriving at what always seems to be the wrong choice. Anticipated negative feeling, remorse even, is part of it. It happens even in the most trivial of situations.

I stand in the supermarket with two tins of peas, one in each hand. I read the labels carefully, noticing the slight difference in price, the brand, the environmental consequences, imagining the colour on the plate, the taste, the texture. I stand in the same spot for several minutes. They're peas for goodness' sake! Can you really suffer from buyer's remorse with peas? I can see some of the other customers looking at me. My hands move up and down, it looks to others as if I might start juggling. I'm weighing the tins of peas up, metaphorically, trying to decide. My inner voice is starting to leak out onto the aisles. I notice that. This is not a good situation. A grey-haired pensioner walks past slowly, using her trolley as support. It squeaks loudly. She stops by the tinned peas. She notices my hand movements; we make eye contact. 'What do you think?' I ask, sensing my opportunity. She looks at me a little surprised, puzzled even. I suspect that she's never been confronted like this before. Maybe she thinks that I work here, doing some sort of market research into peas and that I'd just forgotten my uniform. But she's got time to spare; she pauses. 'The ones on the left,' she says, humouring me, and then grabs a different tin of peas from the shelf and moves on. She glances back as if she wants to get away from me. I feel embarrassed. My dilemma is almost resolved, but not quite. Why did she buy the other peas, if she was that certain and so quick in her decision? I choose the ones that she picked up, go along to the end of the aisle and then stop abruptly, and return to select one of the previous tins, based on price alone. And that night, I notice that they didn't really taste like peas in the end, and I feel myself punished again. My partner stares at me; it's all in that look. Why did you buy *those* peas?

Doubt has all these components – an awareness of the situation and the underlying dilemma, strong affect and sometimes a distinct emotional foreboding, hesitation and in public places embarrassment – regret and anticipated regret, a certain lack of *fluidity* in many situations. A disruption in life. I want to know why some people do not suffer from it when it comes to monumental decisions, and yet there I stand weighing up the pros and cons of tins of peas in public.

So, where do my personal doubts and self-doubts come from? This has puzzled me for many years. I didn't seem to inherit it or learn it from my family in some form of imitative learning. My family were always very decisive; indeed, my brother was very impulsive and confident, they suffered from few doubts (although his impulsivity eventually led him to dying in a climbing accident in the Himalayas, which may have been significant for me certainly later in my life – he was thirty-years old when he died, I was twenty-six), and they were astounded by my early indecision which seemed to get worse as I went through secondary school. I became a doubter. After all, deliberating like this in a supermarket over a tin of peas can have psychological consequences. When you find yourself unable to decide, and the doubt arises, and you become overly aware of that prolonged state of hesitation and prevarication, what happens when it comes to your next important decision? Do you repeat the process? Do you need to go through the same process to deal with the uncertainty and anxiety? Do you become more comfortable with it, no matter how it looks to outsiders? Does it become a coping mechanism? Do you allow it to become a habit? Does it become part of how you see yourself? Does it become part of your identity?

It certainly has for me, and because of that I'm extremely interested in the process. I do seem to suffer from severe doubt that has generalised and spread. I have doubts about myself and my ability to do the simplest things, like changing a fuse, opening a tin can, or changing the oil in my car (my father was a motor mechanic, so the last one is quite odd). I stand helplessly waiting for someone to help. I have doubts about my ability to make the right decisions. Other psychologists might want to label this (prematurely in my mind) as just low self-esteem or just low *self-efficacy* a person's belief or lack of belief that they can be successful when carrying out a particular task. They might say that this should be the starting point of any psychological quest. Just try to understand why you feel so inadequate, they might say, and so ineffectual. But I don't think that I have low self-efficacy about many things – I don't think that I won't succeed when I carry out an academic task, for instance, although strangely doubts creep in when I haven't been tested at this for a period. I need to be tested regularly, and that then makes the doubt disappear temporarily. I prayed a lot about passing exams when I was at school, that too may be important. It was almost superstitious behaviour. It didn't matter what neighbours told me about my ability, or what my grades were at my posh grammar school with a Royal in the title ('that's a great school,' my mother would say, 'they know what they're doing – if you weren't as clever as the nobs, they'd let you know alright'). The doubts would still arise. My inner voice would remind me of how long it was since I passed my last examination. That was always the focus – the gap since I was last tested. Six months is a very long time for the maturing brain, my inner voice would say, your brain may have stopped developing just after that last exam. It might have stopped working; it might have seized up. I worked harder because of that doubt. It was an uncomfortable feeling, but it may have been adaptive in this instance. Many people would describe me as a 'workaholic'. Seeking reassurance through regular effort. Workaholics sometimes seem to achieve a lot; I want to understand how important doubt is as the propellant for all this activity.

Doubt in a variety of forms remains to be uncovered, explored and charted – where doubts come from, how they develop, what they do, how they encourage us to grow and develop or alternatively how they might suppress or destroy us, how they can be used against us.

I have always seen doubt as an affliction, a very personal affliction that needs to be explained somehow. Like most people, psychologists and the public alike, when we suffer from something that we see as an undesirable characteristic, we search for early manifestations and possible influences. But to understand these influences, you need some context. Psychological analyses often tend to very abstract with little connection to individuals and their lives – with some notable exceptions, of course (although the 'Rat Man' may be a little too exotic for many people's taste). The analyses are too often independent of the individual person, but they do sometimes need to be rooted in a description of the circumstances and the life of the individual – they need context and an idiographic rather than nomothetic approach. I will hypothesise about the origins of doubt within certain individuals, and why some people suffer unduly from it, and some *apparently* not at all. I say 'apparently' because it may still have been there in the shadows. Like a forensic detective I will peer behind the curtains to see what clues I can find. Doubt, after all, can be displayed in action, and hesitation, as well as in articulated thoughts which may be concealed from us. I need to locate doubt, and its presence (or apparent absence) within the lives of individuals, including some famous people – great thinkers, artists and scientists. None of us are immune to doubt it seems. But how did doubt fit into their lives? How did they master it? Why was Picasso so supremely confident? The painter who signed his portrait 'I the King'. Why did doubt not inhibit him? Or did it? Was it there somewhere in the shadows? Or Alan Turing? How did he overcome doubts about his ability, and at what cost? Jung explicitly suffered from doubt, but in his view 'suffering' might not be quite the right word. He thought that it was an indispensable component of a complete life. Why was that? Is doubt of the restricting kind all down to overbearing parents with dismissive gestures, and contradictory or implied messages that undermine sense of self or are there other origins, indeed multiple alternative origins? And what are they and how do they affect us? And is absence of doubt all down to the earliest childhood experiences of love, support and affirmation or lack thereof, as Kafka suggested, or is it more complex than that? And how do some people keep doubt in check, whereas others let it take over their lives, their whole lives?

Doubt is more than a symptom of something. It is very much alive, and part of us. It needs to be considered, analysed and understood. It can shape lives. But is doubt always bad? Or is Jung correct? There are many great doubters in religion, history and culture, some good, some not so good, perhaps I'd simply joined them. There's Doubting Thomas for one. They used to snarl his name in our church because of his lack of faith. But I thought that it was a little unfair, why is blind faith so good? He just wanted to see the evidence for himself in the form of the wounds on Christ's body before he could believe in the resurrection. Or the dithering Grand Old Duke of York, plagued by doubt in the form of procrastination, marching his men to the top of the hill, and marching then down again. The Grand Old Duke of York was reputedly Prince Frederick, the Duke of York and Albany, who procrastinated against the French in the Napoleonic Wars and then retreated (although the hill in the town of Cassel isn't much of a hill in very flat Flanders). Then there's René Descartes, known to most for his maxim *cogito ergo sum* ('I think, therefore I am'). He argued that if he doubted, then something or someone must be doing the doubting, which proves his existence – in other words, doubt is an important, indeed essential human process (and a critical method for philosophy). In the case of Descartes and Doubting Thomas doubting would seem to be a good thing; direct empirical evidence is crucially important for science, understanding and even belief. But doubt can also be a bad thing, an obstacle, the doubt that leads to procrastination and delay as in the case of the Grand Old Duke. The doubt that reminds you in your everyday life of future rejection and remorse, the doubt that inhibits, the doubt that prevents you reaching your full potential. This acts as a reminder that there are different kinds of doubt with different kinds of developmental history. I unfortunately am a bit of a mixture. I procrastinate about even very simple things, this marks me out, but these days I like to think that I also use doubt to challenge orthodoxy in different areas of my academic work.

But can we change our doubting habits? We live in this great therapeutic age where everything, it seems, can be modified, if you just want it enough and have access to the right interventions, the right role models and incentives. What role models are out there, and what can we learn from those who have not been inhibited by self-doubt? I watched one celebrated boxing trainer remove doubt from his fighters with the most unconventional of practices. He made them sing nursery rhymes in front of their fellow boxers, and practice forward flips over the ropes into the ring. It seemed to be effective – he removed doubt from a whole gym full of confident 'cocky' fighters and managed to produce one of the most arrogant and doubt-free world boxing champions of all time (Motormouth, as Naseem Hamed was known, would certainly have given Muhammed Ali a run for his money).

But how do those who achieve great things in their life deal with their doubts without 'therapeutic' interventions like this? There is a very long list to choose from – in the arts, in science and literature, in many of the great men and women who have transformed all our lives. Did they manage to achieve all that they did because they had no self-doubt? Where did their confidence in their ability come from? How did they manage any slight doubts? What are the downsides of an absence of doubt, if any? Do we need doubt? Or is it like anxiety, stress and depression that we need to rid ourselves off through individual-based therapy (even very unconventional 'therapy' like that of the boxing trainer) – some form of cognitive behavioural therapy for a better and happier life? Indeed, if we teach people to deal with depression, stress and anxiety, will it just obliterate doubt? Is doubt just some sort of epiphenomenon that sits on the shoulders of these mental ailments to give us something to occupy our time, while these other powerful stressors do their harm. And what about doubt in things like climate change or Covid-19? What doubts drive belief that climate change is a Chinese hoax to cripple the American economy?

Or what about the opposite – people who seem immune to doubt? People who clearly should have doubts but don't – like George W. Bush before the invasion of Iraq with no doubt about Saddam Hussein's 'weapons of mass destruction' or the righteousness of his own cause (Suskind 2004). Or people who report close encounters with aliens (but otherwise seem quite normal). Why don't they have serious doubts about what has happened to them? A number of years ago, I found myself standing in a council flat in Sheffield talking to a very ordinary couple who had not one but several encounters with aliens. I wanted to hear their detailed account – I was interested in any indicators of doubt and uncertainty in how they talked about it, in their hesitations in speech (Beattie 1983), their smirks, their slips and nonverbal leakage (Beattie 2016), where they might reveal that they just enjoyed telling tall tales to gullible strangers like me. But I must say that I saw and heard no sign of any of these. It was all so matter-of-fact – like a visit to the shops, which it was in a way. This couple seemed to be very normal, indeed painfully normal, on any criterion except for their alien encounters. There was no hysteria in their account, no real excitement, and no sign of any obvious psychological disturbance. It was like an everyday experience told in a matter-of-fact way without any uncertainty or doubt.

'The first time I saw aliens,' the wife Jean explained nonchalantly, whilst sitting in her front room,

> was in 1979 in Gleadless in Sheffield [a working-class district of Sheffield] when I was on my way up to the chip shop. I was going up for two fish suppers at about nine o'clock at night and I bumped into a neighbour of mine, Ken and we just stood chatting for a bit. Then I just happened to look up into the sky and there it was. The spacecraft was a saucer shape, like they are. . . . There were two people inside – beautiful looking people. Both had blond hair and very pale complexions. I could see right into the spaceship. It had neon lights just like my kitchen. After a couple of seconds, the UFO just vanished.

She glanced at me to check how the account was going down. 'Oh,' is all I could think of to say, trying not to leak too much information. She poured me another cup of tea.

'But that was only the first time,' continued Jean enthusiastically.

> The aliens returned a few months later and came right into my house this time when I was ill with cancer. They knew, you see. I felt this hand go

onto the top of my head, and I felt these vibrations. They were curing me – they were reassuring me, curing my fear of cancer. When I went into hospital, I had no fear. The doctors couldn't believe how well I was coping with it. They were amazed. I couldn't tell them that it was because of the aliens.

Her husband Tom nodded away in the background, backing up her every word, without saying anything. Tom himself had no direct experience of aliens, and he clearly felt that he was missing out. I could see it in his face. But then fate intervened, and he was given his chance. He said that he saw a spacecraft through some trees in Clumber Park when he and his wife were driving down to Skegness for the weekend. Again, he told the story in a flat, matter-of-fact way. 'The spacecraft was only fifty feet away. I saw the lot. It was black and shiny, just as if it had come out of a car wash.'

But Jean wouldn't let him stop near the spacecraft. This was something of a sore point between them and he's never quite forgiven her for it. 'It was an opportunity of a lifetime that I missed there,' he said wistfully. 'An opportunity of a lifetime,' he repeated. He threw Jean a look. It was as close to a tiff as anything I had seen that day.

'I hope they come back,' Tom said, staring straight at me defiantly. 'I'd go with them.'

Jean said that he should be more careful 'because they might want to experiment on you'.

'I don't care,' said Tom. 'What's there to lose in the end.'

The problem was that I don't think that she wanted to share her alien experiences with him; I could sense it. She wanted to feel special. It felt a bit like there had been some emotional 'infidelity' going on in the background (with the aliens that is). 'But nothing too physical,' she said reassuringly. But she seemed to be thinking of all those good vibrations.

I, of course, had one obvious question for them both. I asked them directly whether they had any doubts about what they had experienced. Could they perhaps have been mistaken? Could it have been some sort of dream or delusional fantasy? I elongated 'fantasy' as if I didn't really want to say it or finish the word. They were not looking pleased.

If aliens are so common in Sheffield, I continued, why don't others report sightings as well. There was no pause, no hesitation. Jean was straight in – she had a ready-made answer to this one, she seemed to be in charge here. She was both fluent and confident.

A lot more people than you think have seen aliens, but they don't want to talk about it. Take Ken, for example, because he had learning difficulties, he thought that people would just laugh at him, so he kept quiet.

Tom just nodded quietly and then made a blowing noise, as if the truth is hard to bear and even harder to hold in.

'And what attracts the aliens to Sheffield in the first place?' I asked. 'Why are the skies above Sheffield filled with alien spacecraft? Why Sheffield?' I was at my cynical best.

Jean had a theory, not a tentative theory nor a guess nor some speculative hypothesis, but one that made great sense to her and that precluded any doubt.

> They probably use the Pennines to guide them in their travels when they're visiting the United Kingdom, the Pennines run right down the middle of the country, which is very handy when you're heading south. And then the aliens just see the lights of Sheffield on the left-hand side, and they decide to pop in for a closer look. They're just curious like everybody else. If you'd been travelling through the darkness of space for months on end and you saw the lights of Sheffield, all lit up, wouldn't you not want to go and have a bit of a gander?

There was a logic to it, or perhaps more accurately a partial logic, based on the assumption that the aliens were nosy buggers just like the rest of us. They may have had their mission in the darkness of endless space, but they could be distracted by a few streetlights and Sheffield market all lit up at night.

Jean sat silent after telling me these secrets about our extraterrestrial friends, and how they navigate and how they think, and how their attention is limited like everyone else's. She had made the supernatural mundane and ordinary. And doubt itself was seemingly eradicated in her account by rooting the supernatural experience of the visiting aliens in the mundane details of everyday life (Beattie 1986; Wooffitt 1992), a narrative constructed to normalise the couple, with trips to the chip shop and weekends in Skegness, and ready explanations as to why other witnesses like Ken would not come forward (Beattie 1998a). And any doubt, of course, by either her or her husband, would destroy the account – it would become a fantasy, a tall tale, a story rather than a reliable account of an actual experience. Instead, it was rock solid and frighteningly mundane. But a lot of work must go into this. It must be that mundane to signify a normal and very average life, a life not full of fanciful ideas or thoughts. I was told that the aliens had the same lights in their spacecraft that Jean had in her own kitchen. But these mundane details are essential to build an account of 'what really happened' and for Jean to represent herself as a genuine witness (Potter 1996) – she had seen the lights in the spacecraft so clearly that she could describe them, indeed recognise them. She presumably would know that this would sound ridiculous, but she risks this dilemma of ridicule for the sake of detail. Any genuine account needs detail. This was a modern-day religious miracle, with aliens instead of angels, in one of the poorest council estates in Sheffield – aliens had come to cure the sick and take the fear of death away. Like all miracles, it needed specifics.

I found the story odd and their certainty about what had happened to them odder still. I wanted that old nagging doubt to appear in this story in order to connect with them as fellow human beings confronted with odd perceptions, and 'something' that may or may not have been out of the ordinary. How can anyone not have doubt that Sheffield is the capital of alien visitations, and that alien spacecraft have the same lights that you can get in the market in Sheffield? What psychological processes underpinned this? Cognitive dissonance might well have had some role to play here. Leon Festinger developed his classic theory of cognitive dissonance in the 1950s to explain what happens when cognitions (opinions, beliefs, knowledge) and knowledge of one's own actions and feelings conflict with each other. This conflict sets up a state of cognitive dissonance characterised by a degree of discomfort that people try to resolve (Festinger 1957). Sightings of aliens outside a chip shop in Sheffield reported to friends and neighbours are likely to produce some discomfort (certainly initially), and Festinger suggested that when we are in a state of cognitive dissonance like this, we find ways of reducing the discomfort either by changing our underlying beliefs or opinions ('it never happened'), or by acquiring new information to make the beliefs more plausible (Jean had been reading about other alien visitations, mainly in the United States) and proselytising about the experience. In Festinger's words – 'If more and more people can be persuaded that the system of belief is correct, then clearly it must, after all, be correct.' Jean had become *more* committed to her belief about what had happened to her through repeated telling of the story. It was also now part of who she was, part of her self-identity.

I am also reminded of Festinger's ethnographic research on a doomsday cult in Chicago as the cult waited for the end of the world in a great flood scheduled, according to the cult's extraterrestrial messengers ('The Guardians') for the 21st December 1954 (Festinger et al. 1956). Festinger described how this cult reacted when the prophecy failed (as the world from their point of view unfortunately didn't end). Many members of the cult had given up their secure jobs and families to join – they were fully committed to their beliefs. In Festinger's words:

> The dissonance is too important and though they may try to hide it, even from themselves, the believers still know that the prediction was false, and all their preparations were in vain. The dissonance cannot be eliminated completely by denying or rationalizing the disconfirmation. But there is a way in which the remaining dissonance can be reduced. If more and more people can be persuaded that the system of belief is correct, then clearly it must, after all, be correct. Consider the extreme case: if everyone in the whole world believed something there would be no question at all as to the validity of this belief. It is for this reason that we observe the increase in proselyting following disconfirmation. If the proselyting proves successful, then by gathering more adherents and effectively surrounding himself with supporters, the believer reduces dissonance to the point where he can live with it.
>
> (Festinger et al. 1956: 27–28)

This might explain why Jean and Tom were so keen to talk about their experiences with aliens to everybody, including cynical researchers like myself. It was a way of dealing with cognitive dissonance and the disappointment (after all the aliens never came back to see them in Gleadless after the first few visits), and to keep the beliefs in place and doubt at bay. Many local people were persuaded of their story because of the level of detail and mundane, ordinariness of the whole thing (but some were not) – they were 'gathering adherents and effectively surrounding [themselves] with supporters', in Festinger's words. So cognitive dissonance may be one mechanism that operates when doubts may arise, but there may be other ways of dealing with it, as we will see.

But their accounts of their experiences with aliens also remind us that there are clearly marked individual differences in the experience and expression of doubt and that the narratives we tell about our lives and our perceptions can act as buffers to the world, and that convincing narratives may end up helping to convince the person generating the narrative as much or more than the audience itself through the reduction of dissonance (Beattie 2018a). From this perspective language use and thinking are intimately connected. We persuade ourselves as well as others, and language has a critical role to play in this.

I wanted to move beyond words at the end of my interview with Jean and Tom. As I left, Jean drew the spacecraft for me – at my request. It was an oval shape with two drivers. Both had familiar haircuts. There was a light on the top. It could have been done by a five-year-old.

'Is this it?' I asked.

'That's it,' said Tom.

Jean looked pleased with her efforts. She was spreading the word to a cynical world. A world full of doubt. My world.

This book must be a journey, and a very personal journey at that. I need to dig below the surface, doubt and doubting may not be obvious on the surface. It's not with me. I've learned to cover up my doubting to colleagues and friends, not like when I was a boy, but it occasionally slips out, like all great unconscious processes, or processes not fully subject to conscious control, like in the supermarket that day with the tins of peas. And then when I stand there, frozen by doubt, weighing up the alternatives, people often think that I'm joking or performing for their amusement. They think that it's a play for their attention. I've learned to disguise my doubting, if not to defeat it.

Doubt is a thing that people rarely talk about. I want to bring it into the open. I want to understand it, where it comes from and what it does. And ultimately, whether it's constructive or destructive. And if it is destructive, what it destroys, apart from our confidence and our ability to connect to others. I want to understand how doubt develops and what feeds it, how it's bound up with other psychological anxieties and insecurities. I want to understand whether it's useful, and if so for what, or whether it just inhibits and distracts. I want to understand how some people overcome their doubts or help others to. I want to understand doubts about things which have very serious consequences, like why some people, including former presidents of the United States, doubt climate change or the nature of the Covid-19 pandemic when, seemingly, the evidence is there right in front of them. I want to understand why, for years, smokers doubted the medical evidence about the deadly effects of smoking, and what makes some politicians so sure that they're right with no apparent doubt about their position or their beliefs. I want to know how doubt inhibited great thinkers, how they dealt with them or lived with them, or whether they were the spur that they needed to push them on to greatness. I explore how doubt can be used by the merchants of doubt out there in the commercial world using subtle and not so subtle techniques to manipulate us. I want to understand why so many have doubts about climate change despite all the scientific evidence to the contrary.

The journey will be important to me. I want to understand how doubt sits within an individual, what reinforces it or keeps it at bay. I want to explore this private world of doubt both in myself and in others. I want to see it for what it is.

- Doubt is a lack of confidence or uncertainty about something or someone, including the self.
- Doubt is central to science, the law, ethics, politics and philosophy which all involve elaborate and often refined adversarial processes to promote, consider and evaluate doubt in the light of the available evidence.
- Doubt is also central to the self.
- Doubt can be a safeguarding mechanism or a distraction.
- Doubt can be rational or irrational, systematic or random, healthy or pathological.
- Doubt is an instrument of rational thought – in science, the law, philosophy and everyday thinking.
- Doubt has been viewed by some (including Freud) as a symptom of major psychological dysfunction, indeed an enduring form of neurosis rooted in the earliest stages of development.
- Obsessive-compulsive disorder is sometimes referred to as a 'disease of doubt'.
- Doubt is a great driver, perhaps the greatest driver of all, *and* a great inhibitor when it comes to human action.
- Doubt drives science, judicial decisions, philosophical understanding, progress, change, positive action but it can also inhibit decision-making and stop change and lead to procrastination, worry, superstition and delay.
- Doubt is internal and conscious ('a perception of indecision') and therefore highly personal.
- Jung wrote that 'Doubt and insecurity are indispensable components of a complete life.'
- Franz Kafka suffered from extreme self-doubt and believed that his overbearing father was the cause of this.

- Doubt can spread.
- Kafka had doubts about his own body and this seriously affected his health.
- There seems to a wide variation in doubting.
- Some people don't seem to have any doubts, including those who have met aliens more than once.
- There seems to be ways of keeping doubt at bay.
- Cognitive dissonance may play some role in this, but there may well be other techniques.

2 Alone in doubt

It was late spring, but it felt like autumn. I was sitting alone on a mountain top just outside Belfast in the pouring rain, looking down onto the city; the rain was driving at me over my left shoulder. I was soaked through.

I had walked up the hill from my home in Ligoniel, a mill village with rows of terraced streets that sits on the outskirts of the city to the north. It was a poor neighbourhood; even the city council called it a slum (but never to our faces) – condemned houses, boys playing in the street without shoes, not enough shoes to go around the large families with sometimes twelve living in a mill house with two bedrooms, a tin bath in the yard but rusted and useless, white bread dipped in white sugar for breakfast. The village was dominated by its linen mills, spinning mills, bleach mills and dams. My family had worked for generations in these mills: my grandfather was a labourer then a rougher, my mother worked in the spinning room, my grandmother and my Aunt Agnes both worked in the carding room where the flax is combed and prepared for spinning and the dry dust from the flax, the pouce, hangs in the air and gets into your lungs. You could always tell the women and the girls from the carding room, my mother always said, by their shortage of breath and their bad cough. Many of them also smoked because they'd been told that it helps clear your lungs. My Aunt Agnes had the worst cough I've ever heard, it bent her double. She died prematurely, of course. I never heard her get through a single day without that cough. And when she died, I never heard a house go so quiet.

I was sitting with a sheet of flapping A4 paper on my lap, as if I might be signalling. Or rather the white sheet was flapping until it became sodden. Now it just lay there. Perhaps, the message had got through. I was trying to write words and phrases into two long columns separated by a long line down the middle of the page. I was using a red pen. Red signals importance, danger; it was reminding me to stop and think. Subconsciously, that is. That is what I was being forced to do. Think about your decision, whilst pressing the page against my wet leg to try to write on it. The page was so damp it started to tear; the pen pushed through the page.

I could see the city stretched out below me; not such a big place from here, given all that it was going to have to go through in the years ahead. You just knew that serious trouble was brewing, it was everywhere, the gossip of the

DOI: 10.4324/9781003282051-2

women on the street, the local news, the men coming home from work, shrugging instead of smiling when confronted with that ubiquitous greeting 'How's it going, mate?' It wasn't the usual answer. I'd never seen so many furrowed brows, tensed eyes, worry, half-suppressed facial expressions, but never fully suppressed – I know that's not possible.

It had started already – the trouble that is. Gusty Spence was in the news, the first terrorist of the Troubles, the first killer. He had re-established the Protestant Ulster Volunteer Force, the UVF that had done such heroic work as the 36th Ulster Division of the British Army on that first day of the Somme on the 1st July 1916 in the First World War. This new UVF had petrol bombed a Catholic-owned pub on the Shankill Road and burnt alive the Protestant pensioner next door – by accident, of course; then a young Catholic man leaving the Malvern Arms on the lower Shankill was shot dead by one of Gusty's gang. This was hardly the UVF famed for going over the top on that fateful morning in July and advancing towards Schwaben Redoubt at the village of Thiepval with their cries of 'No Surrender', with no man turning back. This UVF was different and Gusty Spence, now in prison, was already a folk hero up the Shankill and up our way, and his name was chalked up on our gable wall. The UVF had been born again, which is after all a very Christian notion, but differently.

The trouble had started alright, not the Troubles with that capital letter just yet. It wasn't what had happened, the murders themselves, it was the odd acceptance from many people from this law-abiding and seemingly God-fearing community. It was people's willingness to accept the flimsiest of pretexts for the murders, and then for some to celebrate the murder of these innocents from the other side with paint and spray. That was the most worrying thing.

Of course, you can't condone cold-blooded murder – you need new narratives. These narratives must be made, and then they must be developed slowly conversation by conversation, until they're just right, until they work.

They were saying on the street that the murder of that Catholic fella out with his pals in that bar on the Shankill was an attack on three IRA men, a combat unit no less.

'Your man had to go,' they were saying. 'Get in first, get them at the planning stage. Nip it in the bud.'

A witness had reported in court that it was Gusty himself who had identified the men at the bar as IRA men. Gusty always denied ever saying this – 'How could I even know that?'

And what about the bombing of that Catholic-owned bar? Why was that bombed? 'Oh, you don't know what went on in the back room,' they were saying. 'You don't know what they were up to.'

And that Protestant pensioner? Burned alive by mistake. 'Oh, terrible, terrible, I know. A casualty of war, collateral damage, awful, but that's war for you.'

And this was just an inkling of what would come to pass, of what would become the norm over the next thirty or forty years. The narratives, the justifications in place for actions yet to take place, the emotional dampening of everyday language, new expressions like 'collateral damage' springing from nowhere. Biased accounts of action and planning (Beattie 1992, 2004).

You quickly realise that most things can be described in ways that function as justifications if you, the receiver of the message, are gullible enough.

It was painful to imagine what might happen next in this small city of ours, sitting in that protective huddle, the outstretched arms of hills on one side, Black Mountain, Divis, Wolf Hill, Cave Hill and Belfast Lough on the other. But there's a menace even in landscapes. There was nowhere to run from that city in the bowl of those hills, the arms as if blocking its escape, if the city were to turn on itself.

There were some carefree sheep and lambs in the field beneath me; I felt like warning them of their own fate. Why did they look so complacent wandering about that field? I say 'field', but it was a field with very few boundaries and the sheep and lambs wandered about aimlessly, and out of view. The television aerial was just behind me; this would bring the news of the conflict in the years to come into everyone's home. Black and white images of a march and an ambush on a bridge by some loud bellowing men in suits and Orange Order sashes dangling down onto their proud chests, and then a little later rioting, and old familiar shouts and taunts, used and reused. Convinced of their own position, men without doubt.

All in the very near future but not just yet.

I had climbed up this mountain from one of those mill streets that would become famous in news reels all over the world in the next year or so. It would look familiar to you, not that you've necessarily seen my street before although you may have – there were shootings and murders at the top and the bottom of the street, and a shooting through the front window of one Catholic-owned house right in the middle of the street two doors up from ours. You may not have seen my street, but you will certainly have seen many identical to it. Terraced housing from the end of the nineteenth century, falling down slowly and quietly before your very eyes. The city council had been promising slum clearance for years, ever since I was a child. Maybe the terrorist bombs to come were all planned? They certainly left large open spaces in their wake. The toilet out in the back yard with the crumbling yard walls for privacy, the rusting bath on a nail hammered into the yard wall with the powdery bricks. The bath was just rusted on to the wall, years past possible use, and a mangle from my grandmother's day, rusted and useless too. Rust and crumbling brick, that's what the houses were made of.

But there was always washing and soapy water on the street. I can smell it now. Boiling water from the geyser, hot soapy bubbles, this special aroma, to me it symbolises pride and a sense of community in these working-class streets of ours. The women out on their hands and knees with these big soapy buckets of water in from of them, washing the step outside the house in that almost daily little ritual, where the step was just an imagined private space rather than any kind of raised area.

'Are you washing your step again, Eileen? You'll be wearing that wee step down if you're not careful.'

'Oh, I see you're out again as well,' my mother would say back. 'They've got to look their best.'

But there never was a raised step to be worn down. I wasn't completely sure as a child whether this was some sort of sophisticated joke. But I don't think so. I think that there was a step in the imagination of the women who washed them. Parlour houses down the road from ours with their wee front parlours had an actual step outside their house, signalling their elevated position in social class, up one small but very significant notch from ours; our wee mill houses didn't. We were the bottom rung of the social ladder, the bottom of the heap. So, I always think that it was an imagined, aspirational sort of step.

Then later the TV would show the street with women out in the middle of them, banging bin lids to warn the neighbours of army house searches, but that wasn't our street, that was the other side, the Catholic side, we never banged bin lids like that. We had too much pride. And kids throwing stones, and army snipers inching along on their hunches in their muddy boots, dirtying those newly washed steps, taking cover right up against the front door. Then explosions in the distance at first, and then down at the bottom of the street when our local bar was blown up and the next one just up a bit. And shooting at nights in call and response patterns. Then a huddle at the end of the street itself, with women in their curlers and headscarves, coats thrown hastily over their shoulders, standing in a tight little group, looking down and shaking their heads at something or somebody lying there in the middle of the road. It would be covered up, you wouldn't be able to see what it was, but there would be blood seeping through an old torn blanket just covering it.

'Get one of them old torn blankets out of the back room, not one of the good ones. For God's sake quick.' And the blood would be spreading from a stain to a patch to a back full. Sometimes the blood would leave a pattern or a blob like a Rorschach, to be read, allowing you to project yourself into it.

Some man in his oily overalls on the way home from work would pull the blanket back and the women would gasp again and shake their heads. All these women would be bound together in this apparently endless voyeuristic grief for three or four decades. Watching but not able to do anything. There would be a lot of shaking of heads to come.

But this was all in the future, the very near future, as it turned out, I already sensed that. I was alright though: I would see to that, that's what I told myself.

I was up there on the hill away from it all, not entirely on my own. I was sitting in the rain holding the paper with one hand and trying to stop my dog Keeper tearing down into that field below, and ripping one or more of those sheep apart, with the other. He was watching their every move. He could be a very crafty dog. He was stalking them, tracking their every movement with his stealthy practiced eyes, and just waiting for me to loosen my grip on his collar. He would track one then another. He was acutely aware of any slight change in pressure. He wasn't actually my dog, he was a neighbour's, she had Parkinson's disease. She couldn't look after him, so I took him out for walks like this. Sometimes walks with a purpose rather than just a wee dander up the hill. I didn't have a lead for him. I just dragged him by the collar when I had to, when he saw a sheep or another dog or a passer-by that he didn't like the look of.

One of my mates said his dog only barked at Catholics, or Fenians, as we called them. He even taught his dog to bark every time he said the word 'Fenian'. It was his party piece. Keeper's owner was a Roman Catholic. It wouldn't have worked with him. You couldn't have taught him that trick.

We all knew that something was going to happen, and we knew that it would be soon, now that Gusty was locked up, and celebrated in graffiti in all our little Protestant ghettos. The students up at Queens with the hair over their collars – 'long hairs', my mother called them although their hair wasn't that long – and their university scarves wrapped tightly around them were talking about civil rights and singing Bob Dylan songs on the black-and-white telly in the corner of our front room. Our second-hand TV was always on the blink, it never worked properly. You had to hit it on the side when you were passing to keep the picture on. If you forgot, you'd be reminded.

'Hit that a wee skite for me,' my mother would always say when I came into the front room, 'I've been up and down four times.'

I would walk over to the TV and slap it hard on the side, almost knocking it off its stand.

'Jesus you've hit that too bloody hard, now the picture's gone off completely. You're so bloody careless.'

Then we would sit for a few minutes in front of the television now with no picture at all but with the sound turned up loud to compensate for the lack of vision, and we'd have to listen to all those songs sang loudly and tunelessly by those student-types sitting in the middle of Botanic Avenue protesting about something or other.

'We shall overcome.' Over and over again, sung in that grating and loud (because the sound was up full) middle-class accent of South Belfast, miles from here.

'Jesus, not that bloody song again,' my mother said, 'Don't they know any other songs?'

'And they're nearly all Protestants singing that bloody song,' she added making a blowing movement and just squinting at the TV, just trying to make out the picture through the blurry lines.

> What are they singing about? What the hell do they mean 'the times they are a changin', they need to wise up – things won't change around here, they never do. They don't know what they're talking about. They should get a job before they start preaching to the rest of us about change. We're Protestant and poor, and the next-door neighbours are Catholic and poor. We all work up in the same mill. Things are only going to change if we get together and rob a bank.

The whole situation felt unsettling. I wasn't one of those university types with that green and black scarf, with the thin red lines like train tracks up the middle. I was never going to be. I could never talk like that; my mother would have disowned me. And I certainly wasn't one of those men in the suits with their Orange Order sashes bobbing about their big fat guts as they ran, with the buttons on their shirts popping open, bellowing at the marchers on that bridge. I had no obvious symbol of any allegiances adorning me, nothing to signal to like-minded individuals to come and join me – to accept me into their group. Quite the opposite, in fact, as I sat on that hillside. My dishevelled state, sitting there in that driving rain would have driven people away. I didn't even have that dismissive cynicism of my mother where she would stick to her guns as she made her way to the mill and say over and over again that nothing was going to change anyway around here and that there was no point, so just ignore those university types telling us how to live, and then we'd all return to making the best of what we'd got, which was nothing. Just some steps to polish, even in the rain. I never understood why you would wash steps in the rain – house-proud to a fault in slums. But they do look better when the step was washed – everybody knew that.

I had my own issues at that time, my own concerns. I wasn't going to be walking to that mill every day. That wasn't going to be my destiny. I was going to make sure of that. That's why I had other things in mind. That page in front of me for a start, with Keeper testing my concentration and my resolve by lunging forward every few minutes, just checking that this wasn't the moment to unleash mayhem on those sheep below, that he'd clearly planned in that intelligent, crafty and quite transparent head of his. You could see in his eyes when he had a devious thought in his head, his excitement would rise, his eyes would shine. But if he did that, we'd both be in trouble. Both fugitives. One slip: I knew that was all it would take. One slip and there would be no way out of here. I would be in that street for life, and I'd be useless in the mill – I'm just not very practical you see. I couldn't keep the machines running. I'd be an embarrassment to my family if I worked there.

I had gone up that mountain with my head full of doubts. I had to make a decision in this great time of political crisis in Northern Ireland. What might that be? Whether I was going to join the Civil Rights Movement or the Orange Order? Gusty's mob, the UVF, or the IRA (heaven forbid: I was after all a Protestant boy)? Or some half-cocked party in the middle, not one thing or the other? No, I was on that mountain to decide what subject to drop going into the fourth form. Would it be Latin or biology? That was it – that was the great decision that required a day in the rain on a mountain top. In this great time of political and civil turmoil that's what troubled me most, that's what preyed on my mind.

That's the thing about doubt, it may be triggered by what seems like trivial things. But this decision wasn't trivial to me. You must be more analytic than that.

I feared that I might make the wrong choice. I belonged in that street down below, the one I could just make out, my family had been reared in it for generations, all mill workers, but I'd passed the Eleven Plus, the first in a generation, and now attended the oldest school in the city, where the other pupils didn't come from a street like mine, whose parents – doctors, head teachers, engineers, bank managers, business owners – men who didn't have to get their hands dirty, in my mother's words, who might advise their children; who might counsel them on their academic decisions. These boys and girls with their neat blazers, and starched white shirts and blazers with a Royal crest on it had all made their decisions with seemingly little hesitancy and great confidence, comparing notes, calling out their responses, like they were at a rugger match, shouting out their allegiances 'I'm for Latin', one shouted. 'Seriously!' said his friend, 'I'm definitely going for biology.' Presumably, the outcome of family discussions in the preceding weeks.

I asked my form master whether I could take my form home with me. He looked a little puzzled but agreed. That's why I was up there in the rain. That's what doubt does, I thought to myself. It makes you stand out; it isolates you; it makes you miserable and cold. It makes you feel alone.

'For God's sake, Beattie,' said the boy who sat in the desk in front of me, when he heard me asking about taking the form home. I didn't know his first name, we weren't close, I just knew his surname, and initials. 'It's straightforward – Latin or biology, arts or science, make your bloody mind up. What are you going to be – an artist or a scientist?'

'Or a bloody mill worker, like his parents', said the boy next to him. And they both laughed, 'either subject might work for him then. Both equally bloody useless and irrelevant when you're in the mill.'

My father didn't work in the mill, but I didn't get the chance to say that; he fixed buses in the Fall Road depot. It wouldn't have helped.

If they could have seen me, just sitting in the rain up at the top of Divis, agonising over it, soaked through with that half wild dog of mine. They would have laughed even more, enjoying the spectacle, and it wasn't even my dog. The scholarship boy who worried about everything. My doubts had crept in, that nagging voice in my head again – persistent, insatiable. What use is Latin compared to biology? What might I do in the future? Go off to Queens and try to acquire that accent that grates on me? Join the Orange Order and protest against the Civil Rights Movement, to hang on to what we'd got, which was nothing. Not an inch! No surrender! Surrender what? We had little to surrender. My grandfather's Orange Order sash lay in the left-hand drawer in the dressing table in the front room. It had been there for years, unused, it smelt very musty; my father was never an Orangeman. He didn't see the point.

What should I base this decision on? That's what troubled me. But I soon doubted my ability to make the decision. Doubt became internalised, it was all down to me; then it generalised to other things. It soon became self-doubt. It started with something specific and then spread. And went deeper. It became part of my identity.

I understood that this decision, Latin or biology, was more important for me than my fellow pupils from middle-class Belfast because with the Troubles brewing, educational opportunity would be my only way out of my street and the mill, and away from the conflict. Several of my close friends from the street who didn't have my opportunity stayed there, trapped, they had nowhere else to go, and as my mother would later say they became 'implicated' in the Troubles. A number were murdered, some committed murder and served the best years of their life in prison. They all lost out.

That's how important my decision was.

I knew that the choice regarding my academic future, Latin or biology, the arts or sciences, was important. I knew that I could take no chances with my education. I had turned up for the Eleven Plus with a handful of pencils, suggesting a level of caution even then, and the lead on my first pencil broke, as I had anticipated.

Every time I heard Gusty's name mentioned before I went up the mountain, I was reminded of the importance of my decision. I had been sitting there for several hours in the rain trying to decide between these two subjects, each hand gesturing, weighing up and balancing the choices. I eventually emerged from the mountain top (although Divis is really a hill), rather like Moses one might say, with a long scroll of paper with long columns of pluses and minuses and the decision at the bottom. It contained all the things that might be relevant to both Latin and biology. My mother had told me that our local doctor, Doctor Nelson, wrote his prescriptions in Latin ('It's definitely Latin, you can't understand a bloody word of it') and that if I ever wanted to be a doctor, although I never once suggested that I did, then I'd have to learn Latin just like him – then I'd be able to speak to him about medical matters. My mother worked in the mill, so she definitely was no expert on Latin.

'How do you know that it's Latin?' I'd ask. 'Because I can't make out a word of it; he definitely doesn't write in English.'

'Perhaps, he's just got bad handwriting,' I replied. 'Or he's a very busy man having to scribble away.'

So, on my Moses-like scroll, there was a lot of crossing out in red biro, including medicine with a line half through it on the pro-Latin column. Biology offered me insights into the order of nature and how the body and the brain worked, but Latin also had a significant list under it, and a certain glamour attached to it – the Roman Empire, the legions, Caesar, 'Veni, Vidi, Vici', the Senate, the Coliseum, fragments of schoolboys' poems about the trenches, 'Dulce et decorum est/Pro patria mori', *Ben Hur*, Charlton Heston, maybe even Doctor Nelson. Latin also offered me lots of long and complicated declensions which I enjoyed reciting in our front room (to assuage my doubt about my academic ability) to my father, who understood not a word of what I was saying but praised me endlessly for doing it error-free and for having such a clever son.

But the Latin recitals were over; my father had passed away a few months before that day on the mountain, falling into a coma with a blood clot on the brain during what was supposed to be a routine investigatory operation. I heard the news in the car park of the Royal Victoria Hospital on a wet and cold Sunday night in February. My Aunt Agnes and I had got the bus to the hospital, my Uncle Terence was supposed to have picked us up, so that we could join my mother there, although as a thirteen-year-old boy I wouldn't have been allowed to see him anyway, as he lay there, unconscious for a week. But, according to my mother, he knew she was there. He lay twiddling with her wedding ring from time to time throughout that long week – conscious but unconscious, in this horrifying nether region.

'He knew I was there, until the very end,' she always said.

All I remember that night in the car park were the words, that short dagger-like sentence, 'Billy's dead', and the shriek of my Aunt Agnes, and where we stood – exactly where we stood, and the black car next to us, and the glow of the hospital entrance in the background in the pouring rain, just to the right of my Uncle Terence's shoulder, and me looking at the oily uneven tarmac with those deep puddles, and refusing to lift my head to see their contorted faces, contorted with this unimaginable grief at the death of my father.

'The best man in the world,' they all said. I couldn't bear it.

I say, 'all I remember', but that's a terrible amount of detail after all those decades. I've become very interested in these images in the brain produced by these sorts of events – these extremely vivid visual memories, and how they connect to how we think and the doubts that we have. I'm interested in the relationship between what we remember and how and why we doubt. Our memories are a big part of us, so are our doubts.

Doctor Nelson called down to our house a few days after my father's death to give my mother some nerve tablets and to talk to her. I was told to go into the back room and wait; the back room with the wallpaper hanging off the damp walls. Doctor Nelson explained to my mother that if my father had regained consciousness after the operation then he would have been a 'vegetable'. That was his exact word.

She came into the back room to pass this message on to me, as soon as the doctor had left. There was no other information about what had happened to my father so unexpectedly in that hospital. Maybe they thought that it was all a working-class family from a mill street could understand, all that we could take in.

'Your daddy would have been a vegetable if he'd come around and you wouldn't have wanted that, would you?' said my mother, trying to act not as the messenger of the doctor but a sort of temporary colleague, the collusion of experts, all the time trying stay in control of her emotions. She was gulping for breath.

'I know you loved your father – I've never met a father and son as close as you two – but you wouldn't have wanted that.'

She passed that message on to me as if it should reassure me, as she too, it seems, had been reassured ever so slightly by it. It wasn't so bad that he had died, that's what she was being told, the alternative would have been much worse. She was inviting me to say that I wouldn't have wanted him to regain consciousness either, to build an alliance between the living not the dead.

But this didn't reassure me. It made me more grief-stricken and angrier. 'What did Doctor Nelson say?' I asked. 'What exactly did he say?' I wanted to know the exact words.

'He said that your daddy would have been a vegetable if he'd come around,' replied my mother choking back the tears. She was trying to emphasise 'vegetable', but the bit after the 'veg' was broken and blurred with a sob that stopped her ever finishing it. I was looking down again, I had stopped wanting to look people in the eyes.

'Are you sure he didn't say "*et fuisset ducuntur inriguae*", if Doctor Nelson is so bloody great at Latin?' The Latin words came out with long gaps between words, as I searched my memory for my Latin vocab, and a strange stress pattern because of these pauses and emotion. I just heard this choppy sentence coming out almost below my breath, like someone else was saying it: I didn't mean to hurt anybody, least of all my mother.

'What are you calling me? Why are you calling me that?' She heard that strong syllable in the middle of '*ducuntur*' and was jumping to her own conclusions about my profane and insulting language. And she rushed back to the front room and sobbed into her hanky, side on to the front window, but visible to the whole street. The widowed woman now on public view without anyone to look after her, to protect her, seemingly alone in her grief, with me the schoolboy in the fancy blazer, useless, just protecting myself – as best I could. I never let her see me cry that day or any day.

His death changed the family and me. My mother still worked in the mill, but we had even less money. I know that I worked harder after he died, still desperate to please him, as desperate as only a thirteen-year-old can feel trying to please a dead silenced father. My mother told me that I was an orphan now and that things had to change. That was the word she used, 'orphan', I was shocked. I thought that an orphan was someone who has lost both parents.

I took this to heart, and the change in me was quite literally instantaneous. I woke up the day after the funeral and decided to live differently. I started that very day cleaning up old paraffin heaters for my neighbours and relatives. I worked all day at this, polishing and shining them until they gleamed, and they paid me for the work. My friend Duck and I started a car-washing business the Saturday after the funeral in that cold, cold February and washed six cars that first Saturday, and then up to twelve then fifteen cars every Saturday morning and into the afternoon. The minister was our first customer. I started a paper round with Duck, and helped the milkman deliver milk in the morning. I saved every penny I earned.

I was frantic for some sense of security and some control over my life. I studied and worked harder, I saved harder – as if this might bring me some security. I filled chocolate boxes with sixpences, and they sat all around the front room. I was very good at marbles played out there on the street, I was the best in the neighbourhood even though I had a very unorthodox style (I threw the marbles rather than shooting with them, you had to go back an extra paving stone if you were a thrower), but I would win all the marbles off the other boys and then sell them back to them. My mother always said that there was a queue at the front door from the boys needing to buy some marbles to play with. My brother was astounded when he came home as to where all this money came from. He never had any money.

It was soon after this that I found myself on that sodden hill in the driving rain. Unsure about my academic direction, unsure about life. Who could advise me on my academic path? Who could reassure me? Nobody I knew spoke Latin, including (I was sure) Doctor Nelson. I had now learned to doubt others. Perhaps, I was concerned that all decisions can have terrible consequences. This wasn't necessarily an overt conscious thought all the time, more an unpleasant feeling, but it could suddenly become conscious, even when I didn't want it too. What would have happened if my father had not gone through with this so-called investigatory procedure in the hospital? What would have happened if he hadn't gone into hospital that night? What would have happened if it had been a different cardiac surgeon? Maybe severe doubt can arise linked to unconnected negative experiences and trauma and then it slips over into other, or even all, aspects of life. I had a decision about academic direction that might be critical to my success at that school, amplified by a generalised anxiety about life. This started me on the process of turning decisions into deliberative acts full of pros and cons, lists, and random mental associations (like the ones with Latin – *Ben Hur*; my father in the front room going through my Latin declensions with me), turning life itself into a halting and disfluent process.

I came down from the drizzly mountain top with just one word in the final line on that sodden sheet of paper. LATIN, in bold red ink which had run with the rain and was now barely legible. It was clearly the wrong choice given that I ended up as a psychologist interested in language who needs to know about the brain and the body, and how strokes impact on psychological functioning and speech, depending upon their location and their proximity to the various speech centres in the left and right hemispheres (Beattie and Ellis 2017). It would have been useful to learn more about strokes and the brain, apart from the fact that they can turn you into a vegetable. The circumstances of my father's death may well have helped turn me into a doubter even with straightforward decisions.

Up there in the rain on Divis Mountain was the first time I became aware of my own doubt about what might appear to many to be a relatively straightforward decision. But, of course, many seemingly straightforward decisions can have important consequences. The consequences of any decision aren't the same for everyone. Perhaps, that's where all my doubts come from – some kind of psychic blow, some sort of trauma, and then here this existential crisis that engulfed my family with my father's death. My family was never going to be the same, maybe not even a family any longer. So, we might need to probe and understand the situation and context of severe doubts when they first arise, to understand why they seem to blight some lives and not others and how they might function, and how they can be kept at bay, and how they can change lives.

- Some people seem to suffer unduly from doubt.
- I am one of them.
- Doubt can disrupt the flow of life; it can draw attention to itself. It can isolate you from your peers.
- It has both an emotional and a cognitive component.
- Doubt is not necessarily bad, sometimes it's essential but sometimes it's not.
- Doubting can become a habit.
- Sometimes, people can recall when they first became aware of severe doubt with what many observers might assume was a straightforward decision.
- But some apparently straightforward decisions can have profound consequences, and your subconscious can recognise this and interrupt the decision-making process.
- Doubt may start here and be influenced by other unsettling factors in life.
- Doubt can spread into other aspects of life.
- Some people seem immune to doubt.
- For doubters like me, this is very puzzling.

3 Jung's dream

I was sitting on a boat in the middle of a lake in Switzerland. The water was quite choppy, surprisingly so, mainly because of the movement of the boat and I'm not the best of sailors. I felt a little nauseous. The boat had a wooden floor that was quite slippery. So, I sat where I was, refusing to move. The boat was nearly empty. A bearded man sitting with his wife smiled over at me. It was a benign and open smile, as if he wanted me to say something, to talk about why I was here, at the very tail end of the summer when all the normal holidaymakers had gone. I think he could guess. We were almost certainly here for the same reason. I imagined him as a professor at some lesser European university, but I didn't try to start a conversation. I looked away from him.

I had a notebook on my lap, it was blank. I wasn't sure what I was expecting to write but nothing was coming to me even though we were now approaching the 'tower' on the banks of the lake, which was now quite visible. It was like a castle of the imagination, a castle drawn by a child, a castle from a fairy tale. It was once the home of the great man who changed so much about how we think about the psyche and dreams and the unconscious. He built it himself from his imagination. Having spent years writing about the unconscious and his own dreams and fantasies, he was still not satisfied. 'Words and paper,' he wrote in his autobiography 'did not seem real enough to me; something more was needed. I had to achieve a kind of representation in stone of my innermost thoughts and of the knowledge I had acquired' (Jung 1961/1973: 250). Jung bought the land for the tower in 1922 in Bollingen on the shores of Lake Zurich. It was old church land which once belonged to the monastery of St. Gall.

Jung explained that he wanted to build the house near the water's edge and intended to build a primitive one-storey dwelling like an African hut with a fire ringed by a few stones in the middle of the hut. 'Primitive huts concretise an idea of wholeness, a familial wholeness in which all sorts of small domestic animals likewise participate' (Jung 1961/1973: 250). He added to this concept with a tower-like annex. But again, he wrote that he felt it incomplete. He needed space where no one else would be allowed in, except with his permission, so he built a second tower. He painted on the walls of this space so that he could express 'all those things which have carried me out of time into seclusion, out of the present into timelessness'. After his wife's death in 1955,

DOI: 10.4324/9781003282051-3

he says that he suddenly realised that the small central section 'which crouched so low, so hidden, was myself!', so he needed an upper storey to represent his 'ego-personality'. To his mind this physical extension represented the 'extension of consciousness achieved in old age'. He was being reborn in stone living in 'modest harmony' with nature. There was no electricity and no running water. 'At Bollingen,' he writes, 'I am in the midst of my true life, I am most deeply myself' (Jung 1961/1973: 252).

That's why I wanted to see it, to view the great man, to experience him, this part creator of our psychological age, timelessly embodied in stone, the facets of his personality frozen in turrets and low crouching rooms. I wanted to be moved by recognition and insight, but I couldn't stop thinking about Shelley's Ozymandias – this was my own nagging doubt. Here the traveller, from an antique land, views 'two vast and trunkless legs of stone' sticking out of the desert sand and a pedestal with an inscription which reads 'My name is Ozymandias, king of kings: Look on my works, ye Mighty, and despair!' and the sand 'boundless and bare' stretching in every direction (Shelley 1826).

The castle just sat there on the lapping shore. The professor started to sob – his wife comforted him. I was unmoved by both the house and the professor, which is unusual for me. I sat there frozen, watching the greyish water lapping against the side of the boat.

I suspected that the house signified more about Jung's great doubts rather than anything else – his doubts about Freud, of course, that drove psychology forward (or backwards into its mystical past) but personal doubts that long preceded those, but which may have been critical to his own psychological development as well as his professional and scientific development. Jung famously wrote in his letters (Jung 1951/1976) that 'Doubt and insecurity are indispensable components of a complete life.' From this psychoanalyst – with a unique perspective on his own personal experiences, including his thoughts and dreams, from the earliest of ages – we may learn something about doubt and how it is experienced. In Jung's autobiography where he documents the minutiae of his inner life including the possible role of unconscious processes, we may gain insights into the origins of doubt, at least from the point of view of the subject himself (rather than attempting to understand it from any external, objective 'scientific' perspective), and how such doubts develop, change and permeate outwards to influence other aspects of a life, perhaps making themselves indispensable components of a complete life in the process.

Jung had begun his illustrious psychiatric career heavily influenced by Freud, and by Breuer and Janet. As a young psychiatrist interested in schizophrenia, he found that Freud's technique of dream analysis and interpretation, as outlined in Freud's *The Interpretation of Dreams*, provided valuable insights into schizophrenic forms of expression. He was particularly taken with the idea of the mechanism of repression, which was, of course, critical to dream interpretation – this was something that he had observed in his own work with schizophrenics when they were presented with various word association tasks. On occasion, he noted that their responses were sometimes unduly slow, and Jung hypothesised that the stimulus word had touched 'a psychic lesion or conflict', but that this link was unconscious, and that the mechanism of repression was at work. He felt that his research on schizophrenia essentially corroborated Freud's line of argument developed from those suffering from a variety of neurotic complaints. It demonstrated the power and generality of Freud's idea.

Jung may have agreed with Freud's articulation of the basic mechanism of repression, but he explained that he never agreed with Freud's hypothesis of the cause of the process. Freud thought that aspects of sexuality, and some sort of sexual trauma, were at the centre of things. Jung doubted this – arguing instead that problems of social adaptation, difficult family circumstances, issues to do with identity, status and prestige were likely to be more significant. He presented his doubts to Freud, 'but he would not grant that factors other than sexuality could be the cause. That was highly unsatisfactory to me,' Jung wrote in his autobiography (Jung 1961/1973: 170). It is an interesting comment, as he hints at something beyond the mere exchange of ideas and theories, something more *interpersonal*. Jung wanted his doubts to be considered, but Freud would not engage on this. That's what seems to have hurt Jung the most, so he retaliated, quite explicitly; he attacked Freud – attempting to damage his reputation in his book by highlighting the perception of Freud and his work at that time. He tells us in his autobiography that although he planned to work in this area of repression with due acknowledgment of Freud's contribution, this was difficult, because he wrote:

Freud was definitely *persona non grata* in the academic world at the time, and any connection with him would have been damaging in scientific circles. "Important people" at most mentioned him surreptitiously, and at congresses he was discussed only in the corridors, never on the floor.

(Jung 1961/1973: 170)

Maybe, we, the readers, should align with these 'important people', and after reflection come to share their view of Freud but at the same time come to appreciate Jung himself, this early career psychiatrist, who was a free thinker and a doubter, who followed his gut instincts on this, uninhibited by what important people thought.

But it's even a little more complicated than that. His next set of doubts were to do with whether he should credit Freud's intellectual contribution to his own research here or whether he should publish his work without mentioning Freud because of how authorities in the psychiatric world might react. This was an intellectual and moral dilemma for him – the doubt was explicit and unpleasant. The doubt here he says came from 'the devil' whispering in his ear, only overcome by the voice of his 'second personality', which he often refers to in his writing – 'If you do a thing like that, as if you had no knowledge of Freud, it would be a piece of trickery. You cannot build your life upon a lie. With that, the question was settled' (1961/1973: 171). He says that he acknowledged Freud's contribution in his paper and subsequently 'fought for him' and indeed publicly chastised others who talked about obsessional neuroses without mentioning Freud. But at some cost, he says. He recounts how he was informed by two distinguished professors that he was risking his career by 'siding' with Freud.

This is a story at so many different levels. A historical narrative of opposing views with enormous significance for the history of psychology, leading to a split in psychoanalysis itself – Jung, a follower, indeed a disciple of Freud, was not convinced by what Freud had to say about the sexual origin of the neuroses, and the role of repression in this process. Jung's view was that they may play some small part in the neuroses, but they cannot possibly be the whole story. We need a different sort of account, he argued. But this is also a story of doubt itself. Firstly, how doubt arises in an academic context – the thought that repression may not always have a sexual origin, and what happens if such doubts are not countenanced when presented to the other, but instead summarily rejected. And how even high-minded academics can stoop so low in their attacks to defend their doubts (but again externalising the attacks to others – 'important people' also thought this about Freud). It's also a narrative of how doubt gives rise to other doubts ('should I credit Freud's work in my own research?') and how that doubt must be personified and externalised (the voice of the devil, no less). Doubt may be an unpleasant experience even for the most cultured of thinkers and is clearly dealt with and tolerated in several different ways.

When this doubt is then resolved – no longer just 'Could repression have a non-sexual origin?', but rather 'Repression has clearly a range of origins', Jung's subsequent attack went way beyond academic critique. He suggests that the whole edifice of Freudian theory was attributable to Freud's own 'subjective prejudices', which require, indeed which demand, in-depth psychoanalysis of Freud by Jung himself.

> There was no mistaking the fact that Freud was emotionally involved in his sexual theory to an extraordinary degree. When he spoke of it, his tone became urgent, almost anxious, and all signs of his normally critical and sceptical manner vanished. A strange, deeply moved expression came over his face, the cause of which I was at a loss to understand.
>
> (Jung 1961/1973: 173)

Freud, 'the first man of real importance' that Jung had ever encountered was being demolished like a great building whose time has come to be levelled to the ground. Jung explains that it was a profound disappointment that all of Freud's probing analyses had only succeeded 'in finding nothing more in the depths of the psyche than the all too familiar and "all-too-human" limitations' (Jung 1961/1973: 189).

The basis for Jung's academic rejection of Freudian theory was not primarily academic in an abstract sense but more person-centred and experiential. Freud's theory and its emphasis on sex and repression simply did not chime with Jung's own experiences. In Jung's autobiography, he describes growing up in a village amongst peasants in rural Switzerland where 'Incest and perversion were no remarkable novelties to me and did not call for any special explanation. . . . That cabbages thrive in dung was something I had always taken for granted' (Jung 1961/1973: 190). Freud, he was suggesting, had spent his life on his hands and knees deep in the dung of the human allotment and couldn't dig his way back out. Jung's put-down is a remarkable example of a certain type of the passive-aggressive style of certain academics. 'It's just that all of those people are city folks who know nothing about nature and the human stable,' Jung wrote in his autobiography (Jung 1961/1973: 190).

Doubt is essential in science; it drives science forward. But that feeling associated with uncertainty may be an intensely unpleasant emotion for some (but not necessarily for all) that helps shapes their response to situations. It is sometimes surprising (to me at least) that so many academic disputes become highly personal, and you often witness academics lashing out not just at the ideas of other academics, but more directly at each other. Jung used personal attacks on the scientist behind the theory as an important weapon in his armoury for this academic battle on the nature of the psyche and role of the unconscious. All of this from a great thinker who put so much emphasis on spirituality and higher values and the desire to move away from the human 'stable', with all the connotations of what lies in there. But what other role did doubt play in Jung's life which may help us understand how it developed here and how it energised him in his attempt to destroy his mentor, Freud, through his aspersions about Freud's prejudice, emotionality and irrationality.

In his obituary of Freud, Jung (1939/2003a) wrote, 'The cultural history of the past fifty years is inseparably bound up with the name of Sigmund Freud, the founder of psychoanalysis, who has just died' (1939/2003a: 49). He sets him up before knocking him down, dismissing him as a 'nerve specialist' and little more:

> By training he was no psychiatrist, no psychologist, and no philosopher. In philosophy he lacked even the most rudimentary elements of education. . . . The fact is of importance in understanding Freud's peculiar views, which are distinguished by an apparently total lack of any philosophical premises.
>
> (Jung 1939/2003a: 50)

Without philosophy, and without any *rudimentary* background in philosophy, he implies, how can one ever hope to understand the nature of the psyche? What was distinctive about Freud's approach, according to Jung, was the focus on the 'Neurotically degenerate psyche, unfolding its secrets with a mixture of reluctance and ill-concealed enjoyment under the critical eye of the doctor' (Jung 1939/2003a: 50). There is something about the expression 'ill-concealed enjoyment' here that is particularly telling, as if Freud's patients were getting a kick out of opening up – these traumatic factors were unconscious in the first place because according to Freud, they were all connected with the domain of sex, but Jung is doubtful that this is a valid account: 'Every specialist who has to do with the neuroses knows on the one hand how suggestible the patients are and, on the other, how unreliable are their reports' (1939/2003a: 51). Jung concludes, 'The theory was therefore treading on slippery and treacherous ground' (1939/2003a: 51). Freud tried to salvage something from this according to Jung with the theory of repression and the concept of infantile sexuality (which 'unleashed a storm of indignation and disgust, first of all in professional circles and then among the educated public') in his classic text *The Interpretation of Dreams* (1900), where he analysed dreams as a vehicle for wish fulfilment. Again, Jung's comments on this approach reveal a great deal about aspects of Jung's character. 'For us young psychiatrists it was a fount of illumination, but for our older colleagues it was an object of mockery' (Jung 1939/2003a: 53). As before, when he discusses Freud's lack of knowledge of even 'rudimentary' philosophy, he dismisses and belittles Freud's achievement with this vague unsubstantiated reference to 'our older colleagues' who mocked the theory with the ambiguous implicit quantification in the phrase – was it *all* their older colleagues who mocked the theory, or *most*, or just *some*? Which older colleagues in particular? Why should they be listened to in this context? Does 'older' necessarily mean 'wiser'? Or more stuck in their ways? But Jung doesn't leave it at that:

> What impressed us young psychiatrists most was neither the technique nor the theory, both of which seemed to us highly controversial, but the fact that anyone should have dared to investigate dreams at all.
>
> (Jung 1939/2003a: 53)

Freud's approach was courageous, Jung granted him that, and it did pave the way for young psychiatrists like Jung to follow in his footsteps – courageous but foolhardy and decidedly wrong. It allowed Jung to offer a more high-minded, philosophically grounded approach with the Platonic concept of innate ideas ('supraordinate and pre-existent to all phenomena'; Jung 1953/2003b: 7) metamorphosing into Jung's concept of archetypes (as part of the collective unconscious). And at the opposite end of the spectrum it allowed Jung to offer a more 'balanced' and 'real' approach to Freud's psychoanalysis, the psychology of 'city folks', with an exclusive focus on middle-class, genteel, Viennese (and primarily Jewish) neurotics, but instead focussing on 'ordinary' people, people generally, with a broader early 'education' deriving from Jung's early, earthier experiences in rural Switzerland (where 'Incest and perversion . . . did not call for any special explanation').

It is a barbed obituary throughout. Jung recognises that 'Doubt alone is the mother of scientific truth' (Jung 1939/2003a: 57), but he doesn't just try use doubt to change or rather destroy Freud's theory; he attempts to destroy Freud the man. And what means might a psychoanalyst try to do that? Using psychoanalysis itself, of course. Jung had always argued that Freud's great journey had started with his teacher Charcot at the Salpêtrière and his discovery that hysterical symptoms were the consequence of certain ideas, affective and unconscious in nature, that had taken possession of the patient's mind, analogous in many ways to the demon in the medieval theory of possession. Jung argued that this was the essential basis for Freudian psychoanalysis, so he projects this concept of trauma through possession onto Freud himself:

> In the course of the personal friendship which bound me to Freud for many years, I was permitted a deep glimpse into the mind of this remarkable man. He was a man possessed by a daemon . . . that took possession of his soul and never let him go. . . . All who had a share in the fate of this great man saw this tragedy working out step by step in his life and increasingly narrowing his horizon.
>
> (Jung1939/2003a: 57)

Freud was the man possessed, and the man who had lost his soul to the demon, that's why he couldn't change and see the brilliance of Jung (from Jung's point of view) – the psychoanalyst whose mission it was to rediscover the innate soul (Jung 1953/2003b), following the great philosophers like Plato, the psychoanalyst who understood the real dual nature of the psyche.

This is doubt as a driver, uneasy but compelling, driving out all obstacles in the way, be they ideational or in more personal and human form. And where did this doubting come from? From his earliest childhood memories and experiences, it appears. It's clear from his autobiography that doubt shaped his early life, as he explains in *Memories, Dreams, Reflections* where he recounts his 'truth' about his personal and psychological development – which, in his words, is essentially 'a story of the self-realisation of the unconscious' (1961/1973: 17). It is an extremely vivid account of dreams and thoughts and choices and doubts throughout a long and distinguished life. He begins with a description of life in the vicarage in Laufen above the Falls of the Rhine in Switzerland. He recalls a fantastical dream which he had when he was three years old which he says has preoccupied him all his life. He was in a meadow when he discovered a 'dark, rectangular, stone-lined hole in the ground'. A stone staircase led downwards leading to an arch with a big heavy brocade curtain. He pushed it aside to see a king's throne from a fairy tale with something standing on it that he thought at first was a huge tree trunk, twelve to fifteen feet high, and almost touching the ceiling. In his words,

> it was of a curious composition: it was made of skin and naked flesh, and on top there was something like a rounded head with no face and no hair. On the very top of the head was a single eye, gazing motionlessly upwards.
>
> (1961/1973: 27)

He was worried that this gross thing would crawl off the throne 'like a worm' and 'creep towards me'. At that moment, he heard his mother's voice shout 'Yes, just look at him. That is the man-eater!' This dream, he says, haunted him for years, it made him terrified of sleeping lest it return. 'Only much later,' he writes, 'did I realise that what I had seen was a phallus, and it was decades before I understood that it was a ritual phallus' (1961/1973: 27).

He puzzles over the fact that in this dream of a three-year-old the phallus was 'anatomically correct'. He explains that it was fifty years before he understood its full symbolic significance as he considered the motif of cannibalism in the symbolism of the Mass (the bread and wine representing the body and blood of Jesus throughout the Christian church). This, Jung says, was the other side of God, not on a throne in the sky, but deep in the ground, 'Which gazed fixedly upwards and fed on human flesh' (1961/1973: 29). He never underestimated the importance of this dream:

> Through this childhood dream I was initiated into the secrets of the earth. What happened then was a kind of burial in the earth, and many years were to pass before I came out again. . . . My intellectual life had its unconscious beginnings at the time.
>
> (Jung 1961/1973: 30)

The vision in this dream set him apart, made him doubt the religious views of his father who was a pastor in the Swiss Reformed Church, and made him the precocious doubter who accepted so little – his father's Christian views, his concept of a benevolent God, the role of sex and repression in the formation of the neuroses, the nature of dreams, the role of experience alone in our psychic development. And yet there is much that is disturbing in the content of the dream and there are several possible explanations for how he remembers it. One plausible hypothesis might be that the memory trace of something in an under-ground passage has been reworked over the years, and the 'tree-like structure' that might crawl has had details added that turned it (over many years) into the anatomically correct phallus. But Jung rejects this idea explicitly – he wants to link the dream and its contents to more primitive innate ideas and images that do not require normal experiences to allow them to develop. It may have been an anatomically correct phallus, but it was a ritual phallus given to, rather than created by, primitive people the world over from time immemorial. An alternative, of course, is that he may have seen an erect penis and been disturbed by it, as any three-year-old would. Indeed, in his autobiography, Jung tells us that his parents were sleeping apart during his early childhood and that he slept in his father's room. What might he have glimpsed in that bedroom as his father lay sleeping? But again, he refuses to contemplate any idea that a giant anatomically correct penis appearing in a dream, with all the correct details intact, was in any way connected to witnessing or repressing sex (coincidentally, it's interesting how much psychic energy he devotes to rejecting the idea of sexuality influenc-ing dreams throughout his long career). No, this dream was a revelation, a sign, an initiation, prophetic in its significance. 'What kind of superior intelligence was at work?' he asked. He continued, 'Today I know that it happened in order to bring the greatest possible amount of light into the darkness' (1961/1973: 300), almost biblical in its tone.

But the important point is that this dream became part of his personal story, defining his difference; he had seen things and although he didn't understand their full significance for many years, he knew that the vision of God that his father preached was not correct. There is more to heaven and earth, and life and death, than that. 'In the dream I went down into the hole in the earth and found something very different on a golden throne, something non-human and underworldly, which gazed fixedly upwards and fed on human flesh' (1961/1973: 29). This was his secret that defined him. At primary school, he carved a little manikin with frock coat, top hat and shiny black boots and made a bed for him in his pencil case and hid it along with a smooth oblong blackish stone from the Rhine in the forbidden attic at the top of the house. This was a physical manifestation of his secret:

> No one could discover my secret and destroy it. I felt safe, and the torment-ing sense of being at odds with myself was gone. In all difficult situations, whenever I had done something wrong or my feelings had been hurt, or when my father's irritability or my mother's invalidism oppressed me, I thought of my carefully bedded down and wrapped up manikin and his smooth, prettily coloured stone.
>
> (Jung 1961/1973: 37)

The figure of the manikin at rest in his bed, his precious stone beside him, undisturbed, perhaps a projection of the troubled mind of this child who may well have witnessed something that reappeared in that dream which frightened him so much. But he reframed the experience – still terrifying but reframed as a vision, a revelation, he was born to be different, to see things that his schol-arly and religious father had never seen, never experienced, never thought. The dream and his interpretation made him doubt his father's Christian faith. 'Consciously, I was religious in the Christian sense, though always with the res-ervation: "But it is not so certain as all that!" or, "What about that thing under the ground?"' And when religious teachings were pumped into him, he stepped back in his doubt and thought, 'Yes, but there is something else, something very secret that people don't know about.'

This dream and his way of allaying the anxiety by making the manikin and hiding it, safe in bed with its precious stone, was in his own words the 'climax and the conclusion of my childhood' (1961/1973: 38). He says that he forgot about this until he was thirty-five and these fragments of memory came back to him 'with pristine clarity'. Then for the first time he could make sense of it, he could understand its full significance. The carving and the stone were not childish attempts to deal with anxiety, the critical dimension wasn't the mani-kin tucked up in bed with a precious stone and hidden safely in the forbidden attic where no one would enter, rather these were 'archaic psychic components which have entered the individual psyche without any direct line of tradition' (1961/1973: 38). The stones were like the 'soul-stones near Arleshem, and the Australian *churingas* . . . the manikin was a *kabir*, wrapped in his little cloak, hidden in the *kista*, and provided with a supply of life-force, the oblong black stone' (1961/1973: 38–39). In other words, without instruction or experience

Jung as a child had been engaged in a variety of unconscious rituals – archaic and without any conscious deliberation, moved by great unknown forces.

We may view him as sensitive and troubled as a young child, trying to deal with the pressures of what seems at times like difficult family circumstances permeated throughout with the Christian theologising of his father, doubt and uncertainty arising from these circumstances. But that's not how Jung viewed himself then or later. An image in a dream, strange and unexplained, but clearly full of sexual symbolism, marked him out as different:

> My entire youth can be understood in terms of this secret. It induced in me an almost unendurable loneliness. My one great achievement during those years was that I resisted the temptation to talk about it with anyone. . . . I am a solitary, because I know things and must hint at things which other people do not know, and usually do not even want to know.
>
> (Jung 1961/1973: 58)

He had a secret, the secret gave rise to doubt, the secret meant that he couldn't accept the Christianity of his father, he doubted this and had to reject it, and he then had to reject him, the purveyor of untruths (just as he did subsequently with Freud, his intellectual father, where again he rejected both the doctrine and the man).

Jung's self-narrative is crystal clear on this. One dream in his third year of life made him. It made him doubt established wisdom and challenge it wherever he found it (be it Christianity or Freudian psychoanalysis), but he had no uncertainties about the meaning and interpretation of what he experienced in that dream. In other words, doubt is something that was used and directed and then controlled by Jung:

> Looking back, I now see how very much my development as a child anticipated future events and paved the way for modes of adaption to my father's religious collapse. . . . Although we human beings have our own personal life, we are yet in large measure the representatives, the victims and promoters of a collective spirit whose years we counted in centuries.
>
> (Jung 1961/1973: 111)

His behaviour, his mode of thinking, was shaped not by that vicarage in Laufen but by the collective spirit of those that had passed on, shaping his carving of the manikin, shaping his choice of stone, seeing something in that giant phallus. His theoretical views instigated by doubt but then developed by holding doubt at bay. Some might have thought that the notion of innate Platonic ideas had had their day when Jung was alive; it was the age of empiricism and empirical science, but Jung breathed new life into them. And doubt, either whispering in his ear, or controlled and weaponised, was never far from the action itself.

- Jung was a revolutionary and disruptive thinker, 'a master physician of the soul' in J.B. Priestley's words, who changed how we might think about the nature of the psyche.
- Doubt about Freudian theory drove his work forward. He was not content with what Freud had to say about the sexual origin of the neuroses, and the role of repression in this process. They may play some small part in the neuroses, he thought, but they cannot possible be the whole story.
- He attacked the theory and then the man.
- Freud was obsessed with sex, in Jung's view. He was also bitter, prejudiced and emotionally unstable.
- This might be one strategy to convince Jung himself that he was right, to allay his doubts about his own theorising.
- This was not the first time that doubt had driven Jung to reject a doctrine and then the purveyors of that doctrine.
- He had a dream aged three which he kept secret for most of his life that profoundly affected him – it was a dream of a giant phallus sitting on a throne, like God in the sky but this king was underground. This dream was a revelation. It revealed the nature of God and our existence.
- He rejected the Christianity of his father, who was a pastor, and the Church itself. He attacked both.
- He made a manikin and hid it with a stone, burying the secret. Later, he came to see the carving of the manikin and this stone as primitive and archaic unconscious rituals passed on to him without deliberate instruction and without the benefit of experience.
- He had no doubts about his interpretation. This was the basis of his theory of archetypes and the collective unconscious with innate knowledge passed on through each generation unconsciously.
- It is sometimes easy to consider doubt as a homogeneous entity with certain properties that are either positive or negative within the individual, but when it comes to Jung doubt was used as a driver to give the initial impetus to his own views but had to be allayed by attacking the individuals and institutions behind these competing views, and then controlled and marshalled to allow his views to develop and mature.
- A huge organic view of the psyche and the reintegration of the soul into philosophical theorising might well not have occurred without this dream when he was three, but seemingly without any doubts about the reliability or details of the memory of the dream ('pristine clarity', he asserts) or its interpretation.
- Doubt, it seems, can arise from a dream – from the unconscious.
- This can change the person and make them a doubter.
- But doubt is always selective to a degree – the content of the dream, its accurate recall and its meaning was beyond doubt, according to Jung.

4   Feeling like a fraud

Kafka spent his early life, in his words, skulking around his house like a bank clerk who had committed fraud, nervously waiting to be caught out and humiliated. In school, he thought that his repeated exam success would draw the attention of the teachers who could see through his thin charade, his pretence, and haul him out in front of the class and unmask him as the most ignorant child in the class – someone who had cheated his way to this apparent success. Either he was a cheat, or he was just lucky when it came to the examinations, guessing what to revise, maybe even stealing the papers for a sneak look. Or perhaps worst of all, this boy had just deviously worked out what the teachers wanted to hear and just parroted that back to them in person and in his examinations. These thoughts preoccupied him; Damocles' sword hung over him. He was always anxious, hesitant, silent in the presence of his father. He says that it was all attributable to his father's 'chronic disapproval' and his father's 'belittling judgments'. Kafka would leave examples of his work with his father, but his writing was ignored, it was met by a stony indifference. He wanted praise, love, encouragement but received none of this. Instead, it was just threats, intimidation and fear. Kafka says that it left unable even to think properly. He felt like an impostor, a fraud, keeping up a pretence as a writer, cultured and erudite, but expecting disgrace at any moment, expecting to be exposed.

Kafka, it seems, is not alone. It turns out that 'buried in the hearts and minds' of many high-achieving individuals is the private sense of being an impostor or a fraud (Kolligian and Sternberg 1991: 308). This is a very private sort of doubt, rarely talked about, but with real psychological consequences that are, in the words of Kolligian and Sternberg, both 'distressing and maladaptive'. We who suffer from it hide it well, but it can be extremely distressing.

Contemporary psychological interest in this phenomenon really began with an article by Pauline Rose Clance and Suzanne Imes published in the journal *Psychotherapy: Theory, Research and Practice* in 1978. This paper derived from a series of observations that they had made of over 150 highly successful *women* – either 'respected professionals in their fields' or 'students recognized for their academic excellence' through their psychotherapy practices, or in groups and university classes with large sets of university students. According to the authors, these women, despite their excellent degrees, scholastic honours, praise

DOI: 10.4324/9781003282051-4

and professional recognition, did not internalise their success and considered themselves to be 'impostors' (Clance and Imes 1978: 241). They put their success down to luck or worse – administrative error. They thought that they had managed to fool those around them and those in authority – like their deans or heads of school. They were worried that they would be caught out at any time. This anxiety about being discovered as 'phony' kept them awake at nights. They thought that they were impostors.

That is the account of the origin of this phenomenon in this classic psychological paper but it's not entirely complete. There is one significant omission in how they came to uncover this phenomenon. The origin of the idea was much more personal than that. On her university website, Pauline Clance comes clean and explains that her first inkling of this phenomenon came from her own personal experience – she herself experienced this phenomenon when she was in graduate school at the University of Kentucky – she later coined the concept to describe her own experiences and her own doubts. She explains:

> I would take an important examination and be very afraid that I had failed. I remembered all I did not know rather than what I did. My friends began to be sick of my worrying, so I kept my doubts to myself. I thought my fears were due to my educational background.

She doubted her own ability but did not speak about it. Only when she began teaching in a prominent liberal arts college in her first academic post did she recognise the same shared experience in students who were referred to her for counselling. Then she noticed it in many of her female colleagues.

This 'prominent liberal arts college' was the prestigious Oberlin College, a small elite college situated in northeast Ohio in the heart of the Midwest. It was a big step up from Lynchburg College where she had completed her undergraduate studies. Oberlin was founded in 1833 by a Presbyterian minister, the Rev. John J. Shipherd and a missionary, Philo P. Stewart – their mission was to 'train teachers and other Christian leaders for the boundless most desolate fields in the West'. It is a college with a formidable history – the oldest coeducational college in the United States and, it seems, the second oldest continuously operating coeducational institute of higher learning in the world – women were admitted to the undergraduate programme in 1837. It was also the first college in America to admit Black students (1835); Mary Jane Patterson graduated with a B.A. in Education in 1862, the first black woman to graduate from any American college. This was an elite college with a mission to improve the status of all minorities, and yet there seemed to be barriers to their advancement, barriers that were now often internal rather than external.

Clance and Imes (1978) write:

> Women professionals in our sample feel over evaluated by colleagues and administrators. One woman professor said, "I'm not good enough to be on the faculty here. Some mistake was made in the selection process." Another,

the chairperson of her department, said, "Obviously I'm in this position because my abilities have been overestimated." Another woman with two master's degrees, a PhD, and numerous publications to her credit considered herself unqualified to teach remedial college classes in her field. In other words, these women find innumerable means of negating any external evidence that contradicts their belief that they are, in reality, unintelligent.

(Clance and Imes 1978: 241)

These women, the authors explain, were on every objective index high achievers but were unable to internalise their success and take credit for their achievements. They would be concerned about new tasks and challenges because each new task could potentially expose them as a fraud (Holmes et al. 1993: 48). Self-declared impostors fear that eventually some significant person will discover that they are indeed intellectual impostors. One of their interviewees stated,

I was convinced that I would be discovered as a phony when I took my comprehensive doctoral examination. I thought the final test had come. In one way, I was somewhat relieved at this prospect because the pretence would finally be over. I was shocked when my chairman told me that my answers were excellent and that my paper was one of the best he had seen in his entire career.

(Clance and Imes 1978: 242)

These women had a distorted self-image and could not internalise success. They had difficulty accepting praise (because they thought it wasn't deserved) and were frightened of others discovering their lack of knowledge and ability. They often preferred low-level or less challenging positions so as not to be exposed and had ways of dismissing accumulating objective evidence of their underlying ability. Interestingly, a number engaged in ritualistic behaviour to ensure success (Holmes et al. 1993: 51) – superstitious rituals that they can control and can be used to change (temporarily at least) the focus of their attention as they seek chance but reassuring contingencies. Clance and O'Toole (1987) commented that many of the women were, in fact, 'ingenious' at negating the objective external evidence that testified to their ability. They went on to describe the features of the impostor phenomenon in 'the typical female client':

1   They experience great doubt or fear with any impending exam or project deadline and question whether they will succeed this time.
2   The phenomenon was more likely to occur with introverts than with extraverts.
3   There was a constant dread of evaluation and being found out.
4   They were terrified of failing and 'looking foolish' in front of their colleagues.
5   They were guilty about any success because they thought that it wasn't 'real'.

6 They found it difficult to deal with positive feedback because they thought that it wasn't justified.

7 Clance and Imes stress that the women do not fall into any one diagnostic category but clinical symptoms most frequently reported are 'generalized anxiety, lack of self-confidence, depression, and frustration related to inability to meet self-imposed standards of achievement' (Clance and Imes 1978: 242).

8 They overestimate the abilities of others whilst simultaneously underestimating themselves.

9 They understand what 'intelligence' means in a particular way (and the myths about intelligence) that works to their own detriment.

10 And finally, they are often the recipient of 'false and non-affirming family messages . . . that contradict others' messages about her competence and her family has, subtly or overtly, refused to recognize her specific assets' (Clance and O'Toole 1987: 4).

Given that Pauline Clance had worked in several coeducational institutions (Georgia State University followed on from Oberlin College) when she wrote the paper, it is surprising that men are rarely if ever mentioned in this classic paper (and never analysed). The sample that formed the basis for her analysis in this paper was, of course, exclusively female and heavily biased in other ways as well, including in terms of race, ethnicity, social class, educational level and psychological profile, with a significant proportion of their sample presenting with psychological issues (other than the impostor phenomenon) that required counselling. The authors explain how the sample comprised

> ninety-five undergraduate women and ten PhD faculty women at a small academically acclaimed private Midwestern co-educational college; 15 undergraduates, 20 graduate students, and 10 faculty members at a large southern urban university; six medical students from northern and southern universities; and 22 professional women in such fields as law, anthropology, nursing, counselling, religious education, social work, occupational therapy, and teaching. They were primarily white middle- to upper-class women between the ages of 20 and 45. Approximately one-third were therapy clients with specific presenting problems (other than the impostor problem); the other two-thirds were in growth-oriented interaction groups or classes taught by the authors.
>
> (Clance and Imes 1978: 242)

Of course, with a sample like this one must be careful about drawing more general conclusions about the incidence of the impostor phenomenon across genders, race, class and so on, as well as its connection with other psychological issues like generalised anxiety. Presumably, many of the students seeking counselling and finding themselves included in the sample presented with generalised anxiety and similar problems.

One must be even more cautious about trying to unravel the aetiology of the impostor phenomenon at a more general level with so biased a sample. But Clance and Imes do suggest several mechanisms. The obvious question is would the same mechanisms underlie its development in men? The authors write:

> Why do so many bright women, despite consistent and impressive evidence to the contrary, continue to see themselves as impostors who pretend to be bright but who really are not? What are the origins and dynamics of such a belief and what functions could be served by holding on to such a belief?
>
> (Clance and Imes 1978: 242)

They say that they observed that their 'impostors' typically fell into one of two groups, with respect to their early family history. In one group were women who had a sibling thought to be the 'intelligent' member of the family. They, on the other hand, were told that they were the 'sensitive' or emotionally intelligent member of the family. This early assignment to respective roles by their parents can have profound implications, they say:

> One part of her believes the family myth; another part wants to disprove it. School gives her an opportunity to try to prove it to her family and herself that she is bright. She succeeds in obtaining outstanding grades, academic honors, and acclaim from teachers. She feels good about her performance and hopes her family will acknowledge that she is more than just sensitive or charming.
>
> (Clance and Imes 1978: 243)

But the family are still unimpressed, continuing to attribute greater intelligence to the 'bright' sibling whose academic performance is often poorer by comparison. So the impostor is driven to find ways of getting validation for her intellectual competence but secretly doubting her intellect (after all, that has been what she has been told repeatedly), and starts to wonder if she has gained those high marks not because of her intelligence but through other means – through sensitivity to teachers' expectations, through her social skills or her 'feminine charms' (of course, implicit in their argument here, and the use of this particular term, is that it is the male sibling who has been told that he is 'naturally' intelligent). A different family dynamic operates for the second group of women experiencing the impostor phenomenon, they argue. Here, the family conveys to the girl that she is superior to other children in every way – not just in terms of intellect but in terms of personality, appearance, and talents. In the words of the authors:

> There is nothing that she cannot do if she wants to, and she can do it with ease. She is told numerous examples of how she demonstrated her precocity as an infant and toddler, such as learning to talk and read very early or reciting nursery rhymes. In the family members' eyes, she is perfect. The

child, however, begins to have experiences in which she cannot do any and everything she wants to. She does have difficulty in achieving certain things. Yet she feels obligated to fulfil expectations of her family, even though she knows she cannot keep up the act forever. Because she is so indiscriminately praised for everything, she begins to distrust her parents' perceptions of her. Moreover, she begins to doubt herself. When she goes to school her doubts about her abilities are intensified. Although she does outstanding work, she does have to study to do well. Having internalized her parents' definition of brightness as "perfection with ease," and realizing that she cannot live up to this standard; she jumps to the conclusion that she must be dumb. She is not a genius; therefore, she must be an intellectual impostor.

(Clance and Imes 1978: 244)

There are a number of obvious comments to be made on these 'observations' of the early family lives of those women who suffer from the impostor phenomenon. The first is that descriptions of the family dynamics are constructed from the narratives of the individuals themselves rather than from any observations of the families in the home or in group therapy. There is no attempt to triangulate the information with other members of the family as a way of reconstructing any elements of the past. Of course, the narratives themselves may be the critical elements in establishing the impostor phenomenon and keeping it in place, but that is not how Clance and Imes frame it. They talk about 'early family history' and 'family dynamics', with expressions like 'she is indiscriminately praised', all couched in objective behavioural terms, as if they had a lens on the past rather than one person's perception and understanding of it. We just need to be careful that at best we are dealing with individuals' experiences, perceptions and versions of the past, but these versions, of course, may be critical (as they undoubtedly were for Kafka and others). It is our lives, as we understand them, that drive us forward and gives rise to doubts, including severe doubts, along the way.

Secondly, although it might seem intuitively plausible (especially in the early 1970s) that these two types of family dynamic might be more pronounced for female members of the family, there is no a priori reason why they might not apply to men as well. I can find some resonance in my own family (as I'm sure many people can). My brother Bill did not seem that gifted academically in objective terms – he failed the Eleven Plus and went to the local secondary school before leaving at sixteen to become an apprentice electrician. But my mother constantly said that he could do 'just as well as our Geoffrey at school' if he just put his mind to it. The inference was that he was naturally very intelligent (perhaps the most intelligent member of the family given that he hadn't had years of training in academic work) but just chose not to do it because, presumably, he had better things to do with his time than sit in the back bedroom of a slum house poring over books. He was out enjoying himself with his friends and his girlfriend whilst I was stuck inside (Beattie 2021). Of course, my mother's comments could be construed simply as a way of encouraging Bill; it might only have been my acute sensitivity that suggested otherwise. Bill told me constantly that I did more homework that anybody else, that was why I did so well. It wasn't brains, it was effort. My mother told me that I would need glasses if I kept studying like this and that (implicitly) I'd never find a girlfriend and have the opportunities that my brother had with the opposite sex. It was only when I went to university, and I could see others studying up close that I found that this was false – I seemed to work fewer hours than my fellow students but with much better results. My brother died young in a climbing accident in the Himalayas (he had abandoned his work as an electrician as soon as he could to become a professional climber), so he never witnessed any subsequent success on my part or any of the doubts that arose along the way.

It is also interesting that Clance and Imes assume that there are two different family dynamics that operate with the impostor phenomenon. The first 'dynamic' is the explicit comparison with a sibling who is more 'naturally intelligent' where this family expression of being less talented than a sibling becomes internalised and success cannot override this. The second is the vote of confidence by the family – that any academic success on the individual's part can be accomplished because of their great ability and with little effort – 'There is nothing that she cannot do if she wants to, and she can do it with ease' in the words of Clance and Imes, and that if a particular project requires a lot of work and effort (as many do) then they must be an impostor because it should be effortless. But one might suggest that these are not mutually exclusive categories, or indeed characterisations of *different family dynamics*, but perhaps different dynamics at different times, or even different family members. They can operate within the same family and the same child. Academic work is, after all, characterised by a certain degree of longevity – from primary school through secondary school and university, and for some right into and through an academic career. There are many opportunities to say different things at different times and make highly varied comments. Even my brother changed his tune towards the end. 'God knows where you got your brains from,' he said. 'Maybe it was our family doctor; he always spent a long time with our mother.' It was just his little joke. But you can imagine these two different family dynamics operating as twin pressures (often within the same family) – having to prove that you are as intelligent or more intelligent than other siblings and having to do it with ease. This might well apply equally to both men and women. It is hard to think of reasons why it should not.

But does it have the same effect on men and women? We need to consider gender differences in attributions at this point, as indeed Clance and Imes did in their article, and differences in so-called attributional style which is how men and women make attributions for successes and failures in their life. If you attribute your success in academic work to internal factors ('I'm smart') that are stable across time and likely to affect other aspects of your life, so-called global attributions (Seligman 2002) then you are making a more enhancing and optimistic set of attributions for the future – especially if you put any failures down to external factors ('it was a very hard exam') that are unstable ('it was just that one bad exam') and specific ('it was just one exam on that subject'). Such an attributional style may make you more resilient when confronted with failure and perhaps more resilient to the vagaries of the comments of family members. There are several reports of gender differences in this regard. For example, Stipek and Gralinski (1991) reported that when it came to mathematics, girls rated their ability lower than boys, expected to do less well, and were less likely than boys to attribute success to high ability (see also Nicholls 1984; Lohbeck et al. 2017). Nicholls (1975) found that different sorts of attributions for success and failure by girls and boys are already evident by the age of ten. But others have reported no significant gender differences in attributions for success and failure (Voyles and Williams 2004; Wilson et al. 2002). What we certainly know is that there are marked *individual* differences in attributional styles and that such differences are not trivial – internal, stable and global attributions for success are associated with optimism and happiness (see Beattie et al. 2017); internal, stable and global attributions for failure are associated with clinical depression, for example (Beattie 2011). But, of course, attributional style is not present at birth, it develops during socialisation experiences, and family dynamics are likely to be one powerful influence. So even if we found significant and consistent gender differences in attributional style between men and women, with men internalising success and making more optimistic attributions, we couldn't necessarily assume that this style had been in place as a potential buffer throughout the full course of development when the 'family dynamics' identified by Clance and Imes are operating. Such dynamics might be helping to establish the underlying patterns of thinking in the first place.

Clance has argued throughout her published work that the impostor phenomenon, even if it is observed in men, impacts on women more but this just seems to be an opinion rather than an observation based on hard evidence. Indeed, she and Imes openly introduce their hypothesis explicitly as an opinion. They are discussing gender differences in response to an invitation to do honours in a private US college in the years 1969–1973 and write 'Why did it impact women more? The *opinion* of Clance and Imes is that the men were encouraged by mentors, faculty, and the society to go ahead and do honors despite their impostor fears. They were encouraged to override their fears and to go for success' (Clance and O'Toole 1987: 2). But this seems to me a little disingenuous. Surely, the women they investigated were encouraged to go for success, despite their inner or outer anxieties, otherwise they would never have made it to the honours programme (let alone the faculty) in Oberlin College or Georgia State University in the first place.

There are clearly shades of Kafka throughout the observations of Clance and Imes. And Kafka, of course, felt that he could localise the source of his self-doubts and his experience of the impostor phenomenon to his father and his treatment of Kafka as a child, in other words from his perspective in the dynamics if the family. But, in Clance and Imes' (1978) research the focus, as we have seen, is exclusively on high-achieving women that has led many to believe that this is a quasi-clinical phenomenon that is exclusive to women, or that particularly involves women. So, is any of this work relevant to Kafka's own experiences? The answer would appear to be yes. In the intervening four decades we know much more about the incidence of the impostor phenomenon across genders. A systematic review of the impostor phenomenon published in 2019, which covered 62 studies involving over 14,000 participants, recorded that thirty-three articles had compared the rates of the impostor phenomenon by gender – sixteen of these studies did find significantly higher rates in women, but seventeen studies found no significant difference (a very high proportion given the tendency of journals not to publish non-significant findings). In other words, it is something that affects men as well, and clearly many men have indeed felt like impostors.

Like Pauline Clance, I can localise my most intense feeling of being an impostor; it was marked by a transition between universities (perhaps analogous to Clance's own journey when she started at Oberlin College). I described this transition in my memoir *Selfless*. I graduated with a First from the University of Birmingham and was offered several PhD places, at Oxford and Cambridge and several other top universities. I chose Cambridge. I then had to apply to a college at Cambridge and I chose Trinity. Trinity is the greatest and most prestigious college at either Oxford or Cambridge. More Nobel Prizes than France (such a Trinity cliché), six British prime ministers (all Tory or Whig), several kings and future kings of England (Edward VII, George VI, Prince Charles), great scientists, artists, philosophers, mathematicians and historians (Sir Isaac Newton, Lord Rutherford, Niels Bohr, Lord Tennyson, Lord Byron, Sir Francis Bacon, Bertrand Russell, Wittgenstein, Charles Babbage, G.H. Hardy, Lord Macaulay, G.M. Trevelyan, E.H. Carr). I never expected to get in. It almost felt like a dare to apply in the first place. My supervisor in the Psychology Department was Dr Brian Butterworth. I remember the interview well. I had bought a tie and a suit. My new white shirt didn't fit me; I thought that I was being strangled. Brian turned up in a leather jacket and I spent several valuable minutes apologising about how I looked and explaining that I didn't normally wear a suit or a tie. I loosened my tie in the middle of a sentence. I looked like a drunk at closing time. I was trying to disguise my working-class background in that suit. Brian's jacket told me that clothes would not work here as any sort of disguise, but they would still work as a symbol of power and authority – those with power and authority can dress how they choose. But he liked the meticulous and slightly obsessive details of my work, if not my clothes, and offered me a place.

Brian had three new PhD students starting that Michaelmas Term, all interested in language – one interested in the philosophy of language, one in language development, and myself-interested in pauses in speech as indicators of cognitive activity to uncover the psycholinguistic processes underlying the generation of spontaneous speech. Pauses indicate doubt. He asked which one of us would like to go first and make a seminar presentation to a new research group that he was setting up. I volunteered to give a talk based on my extended essay, a critique of artificial intelligence as applied to language. After all, it had been awarded the highest mark ever awarded to an essay in psychology at the University of Birmingham. I was told that it was publishable, just like my dissertation. I shone with confidence *temporarily* as I volunteered.

So, I went away to write the talk in my lonely college room and stared out through the sash windows of Burrell's Field, past the poster of the Death of Chatterton by the Pre-Raphaelite painter Henry Wallis, which I'd bought on my first day in Cambridge. This was my one Cambridge affectation. The image of the Romantic hero, the seventeen-year-old English early Romantic poet Thomas Chatterton dead on his bed, from suicide, his precocious talent rejected and despised, killing himself with arsenic. The windows in the painting were a little like Burrell's Field – I had noticed that straight away but that was where the similarity ended – and the anticipated rejection, of course. I was starting to feel that. They didn't have people from my background in Trinity in the 1970s. Rejection was sure to come. But surely, I had *some* talent to be there, that's what I kept trying to tell myself, but serious doubts had started to appear. Cracks in the edifice built so carefully through the encouragement of my tutors at the University of Birmingham. Sometimes you can write a sentence with great confidence but when you have to say it out loud and potentially be questioned on it, it's a different matter. There were a couple of arguments that I had used in the essay that I now wasn't so convinced of – indeed I wasn't entirely sure that I fully grasped the complexity of some of the underlying arguments. Doubts were now everywhere. But I tried to push these negative thoughts to the back of my mind. Surely nobody would ever ask about this.

Our postgraduate offices and Brian's office were in the Low Temperature Building, on the top floor, just opposite the famous Psychological Laboratory on the Downing Site at the University of Cambridge. The seminar was to be held in the seminar room in the main building. I sat there with my talk on my lap handwritten on acetates waiting for the audience, who I assumed would be fellow postgraduate students, to arrive. Brian sat beside me commenting on each individual as they entered the room:

> There's Professor John Morton, you've probably heard of his logogen model – I've got great problems with his model – he knows my objections, of course, we argue about it constantly; there's Dr. Bernard Comrie, very smart, he's a linguist who works on linguistic universals, . . . there's Professor Brian Josephson he's a bit of a character, he was awarded the Nobel Prize in Physics last year for some experimental work that he did when he was a twenty-two-year-old PhD student here at Cambridge. You don't need to worry about him though. He's just an enthusiastic amateur when it comes to psychology, even when it comes to artificial intelligence, although that is a bit closer to physics and he does know a thing or two about that.

Brian laughed politely and quietly at his own joke. The list went on and on. I think that I had started to shake. The other two PhD students sat grinning at me from the front row.

Brian introduced me; all I remember hearing was 'Birmingham'. That's all. Birmingham. My attention had narrowed significantly. I tried to speak. The first word that came out was 'ah', which, of course, isn't a word. It's a filled pause, a hesitation filled with noise, the very thing I was aiming to study. There was a tremble in my voice. I'm sure that the front row was grinning more now. Belfast accents have a sing-song quality at the best of times, with a rising intonation at the ends of sentences, as if we are asking questions all the time, or unsure of what we're saying. I proceeded for about ten minutes, bombarding them with statements that sounded like questions, filled pauses, repetitions, and I'd now started saying 'you know' every other sentence. This is a very working-class Belfast affliction. Then a hand went up from the back of the room, a room that I noticed had suddenly got much warmer. It was on that precise part of the talk which I had recognised beforehand was my weakest point, the area where my knowledge was shakiest, where I had the thinnest slither of understanding. It was the question that I dreaded, it would expose me fully. It was partly the way the question was phrased. The speaker began, 'Surely, you're not saying. . . .'

I hardly listened after that opening. He was clearly implying that only a fool would say what I had just enunciated. I couldn't remember whether he was a psychologist or a linguist, a world expert or an enthusiastic amateur, a Nobel laureate or a technician that had just wandered in, all the names and descriptions of those in the room were now muddled in my head, I was so tense that I could hardly remember anything, I was thinking that maybe Wittgenstein himself had popped along. The truth is that at that precise moment in time (very embarrassingly), I couldn't remember whether Wittgenstein was alive or dead, all I remember being told a few days earlier was that he liked to call into the Arts Theatre in Cambridge in the afternoons and eat a pork pie, and that to me sounded awfully contemporary. Perhaps, it was the ghost of Wittgenstein that was now in the room there to expose my philosophical muddle. I couldn't see the face of the person who had asked the question, perhaps nobody had asked it, perhaps I had imagined it, perhaps it was my conscience talking, calling me out, calling me a fraud.

Somebody jumped in and then another. 'No, of course, he's not saying that. Only a fool. . . .' I wanted to put my fingers in my ears. They were arguing over what I was intending to say, over what I meant. Some attributed great learning and understanding to me, some thought that it was just a fine ironic touch, some thought that I was just being provocative, 'and he's succeeded', one erudite and very refined voice from the back suddenly shouted.

But this was just the start. They then started arguing about the substance, the theory, the underling epistemological assumptions, about grammar, typology, implicature, linguistic universals, Boolean logic, fuzzy logic, neurons, signal detection theory, neural networks, mathematics, physics. Everybody had their say, except me that is, sitting there with nothing to say, shaking, and trying to control the more obvious bodily movements.

After an hour or so (although that is only an estimate) Brian took control again. 'All very interesting,' he said, 'but I'm sure that Geoffrey's got a lot more to say on this. I think that we should let him continue.' And again, all eyes were fixed on me.

My filled pause was even longer this time. It came out like 'aaaaaaaah', as if I had been stabbed in the neck, and I heard myself saying 'I think most of the points that I was going to make have been covered,' and I left it at that. The group then started chatting amongst themselves, the front row still grinned, and Brian seemed to turn his back on me to chat to somebody else. I sat staring straight ahead with my hands clasped tightly, the sweat from my hands was making the non-permanent ink on the acetates run. They looked like the streaks on a woman's face after she's been crying on a night out, the mascara running everywhere.

They all went to the pub afterwards, I went home. No, I didn't go home, I went back to my room in Burrell's Field, my mind was blank. I stared blankly out of the window, too terrified to start up even an internal dialogue.

I went to that seminar group every week for a year and didn't once dare open my mouth again. I know that for a fact because a record was kept of all the questions and answers in the seminar each week and my name was completely absent. Brian pointed this out to me. I felt like a total impostor – I stood out a mile. The others were so clever and knowledgeable, and not just about psychology, but about everything, about things I had no opinion on, about things I had never thought about, about things I had never heard of. I thought that I was out of my depth. I felt that I had worked out what my lecturers at Birmingham had wanted in their essays and exams. It was almost as if I had I deceived them. I was street-smart rather than genuinely clever, and not even as street-smart as the kids from my street who really did know how to survive on the street. I was just more street-smart than those in university. They were too middle class to recognise it. Now it was different; now I had been exposed. It was a public humiliation.

Eventually, Brian called me into his office and explained that if I didn't speak, he would stop inviting me to the group. So, I plucked up the courage at one talk and asked several questions. When I saw the transcript the following week, there was just a question mark where my name should have been. Nobody knew who I was any longer. I'd become anonymous. I also noted that when I was asking those questions, I had developed a very distinct stammer which most of the time was now there.

I had shown in my undergraduate dissertation that you can produce filled hesitation in normally fluent speakers by changing the social contingencies and 'punishing' unfilled pauses through apparent negative evaluation (Beattie and Bradbury 1979). I needed those social contingencies to change. I thought that there was only one way to do this – I needed to learn properly. I went to classes in linguistics and talks in the evening in Trinity on philosophy and literature, I went to seminars at the Chaucer Club in the Applied Psychology Unit. I had been lulled into a false sense of security by what I now recognised as my modest-enough academic achievements so far (and my street-smart ways). I realised that I had a long way to go. I needed more background knowledge and confidence, to hold my own. I had to develop an inner critic for my ideas and realised that I couldn't leave this criticism to others in a social or academic situation. I needed to challenge myself in advance, and if it looked a little odd, me just sitting there, in silence, gesturing away, arguing with myself, back and forward, to and fro, Punch and Judy, then so be it. I might look a little out of place in the Copper Kettle on King's Parade, just sitting there, over a coffee, talking to myself, with an internal dialogue that sometimes leaked out into an embarrassed social space of foreign language students and tourists, but that was the price I was going to have to pay. So, this is Cambridge, I thought to myself.

My mother, however, was now convinced that I found academic work very easy and that it was effortless to me. That was one pressure that Clance and Imes had surely got right. That year it wasn't. I knew that there were enormous gaps in my knowledge and that I had to catch up. I didn't want her to discover the truth. I couldn't talk about my work, or Cambridge, or my life. She said that 'I went awful quiet.' But that's what impostors do. They keep secrets. Some have tried to argue that the impostor phenomenon is not necessarily a bad thing. Gadsby (2021) suggests that there is 'an overlooked benefit of the condition', namely that this experience can drive one to try harder to succeed – 'a belief that one lacks ability will have a motivating (rather than demotivating) effect' in domains like academe where you need considerable effort to succeed (Gadsby 2021: 7). It did for me but clearly not for everyone.

Want and Kleitman (2006) discovered the opposite – they observed a correlation between level of feelings of being an impostor and self-handicapping tendencies, which involve placing an obstacle in the path of an evaluation so that possible failure can be blamed on the obstacle, such as not preparing adequately for an exam, rather than on yourself. That way you can protect your self-image because any failure does not reflect on you and your ability. You can self-destruct with it. Interestingly, this study also found a correlation between lack of *paternal* care and overprotection by the father and levels of the impostor phenomenon in the child. Thirty-seven per cent of their sample were male. The authors concluded that

> The finding that the role of the father may be especially significant in development of impostor feelings is a new and notable addition to the literature on the family background of the impostor. . . . The overprotective father may have had a narcissistic involvement in the child's achievements, and the self-criticism imposed on themselves by impostors may be an internalisation of the parental desire for success.
>
> (Want and Kleitman 2006: 968)

This, of course, takes us right back to Kafka and his letter, but it doesn't help me. I didn't have an overprotective father with a narcissistic involvement in my work through secondary schools; I had a dead father. But his absence may have given rise to other doubts, which generalised to my feelings about myself. In the 1970s the impostor phenomenon seems to have been rediscovered and assumed to mainly involve women. Nobody thought twice about research that excluded men altogether. Now we know that it is much more widespread than that and that it is found across genders, races, social class and ages (see Bravata et al. 2019). We know much less about how and why it develops. Want and Kleitman (2006) finish their article with the admonition that 'Further investigation is also required into the role of fathers in the development of impostor feelings.' I feel that we are still waiting.

Pauline Clance always used the expression the 'impostor *phenomenon*' carefully throughout her research. In 1991, John Kolligian and Robert Sternberg from Yale asked whether there is such a thing as an 'impostor syndrome'. They begin by critiquing the general construct validity of the Impostor Phenomenon Scale (IPS), citing evidence that some researchers (Edwards et al. 1987) had found an unacceptably low level of internal-consistency reliability for the full scale (alpha = 0.34) compared with the higher reliability of 0.75 reported in earlier research. They argued that we need to think more carefully about the components that make up the experience of perceived fraudulence, and they identified several different dispositional factors that may contribute to this. The first is depressive symptomatology – individuals high on perceived fraudulence may be characterised by distorted attribution processes (also identified by Clance and Imes) that relate to depressive cognition, dysphoric affect and low self-esteem. Nevertheless, despite their inability to internalise success, they set high standards for achievement, leading to a constant fear of failure. The second component they suggest is social anxiety related to both evaluative and social situations – those high in perceived fraudulence are especially prone to anxiety about negative outcomes and exposure as a fraud. The third component of perceived fraudulence is high levels of self-consciousness and a preoccupation with the reaction of others. They may think that other people are as concerned with their thoughts and behaviours as they themselves are. Kolligian and Sternberg (1991) say that this can give rise to high impression-management and self-monitoring skills 'designed to shape others' opinions.' The researchers developed a new 51-item Perceived Fraudulence Scale (PFS) with much higher reliability (alpha = 0.94) consisting of two major factors: factor 1, 'inauthenticity' (e.g. 'In some situations I feel like a "great pretender"; that is, I'm not as genuine as others think I am'), and factor 2, 'self-deprecation' ('When I receive a compliment about my academic or professional abilities, I sometimes find myself making excuses for and explaining away the complement'), along with various measures of achievement pressure, as well as depression, self-esteem, self-monitoring, social anxiety, and reports of open-ended thoughts and feelings coded using an adjective checklist and interviews. Factor 1 (inauthenticity) correlated most highly with measures of self-monitoring and with self-critical aspects of depressive symptoms, achievement pressures and social anxiety and correlated negatively with self-esteem. This factor represents an important component of perceived fraudulence, with a critical combination of high self-monitoring or impression management skills and self-critical or dysphoric personality tendencies. This factor also correlated most highly with the subjects' scores on self-perceptions of fraudulence in the interviews. Factor 2 (self-deprecation) correlated most highly with self-critical and dependent aspects of depression and negatively with self-esteem; it also correlated positively with social anxiety. This factor was significantly correlated with negative affect in the open-ended thoughts and feelings.

This was a step forward in understanding the impostor phenomenon. It suggests that perceived fraudulence involves 'a complex interplay of unauthentic ideation, depressive tendencies, self-criticism, social anxiety, high self-monitoring skills and strong pressures to excel and to achieve. . . . maintained at a high clinical cost' (Kolligian and Sternberg 1991: 323). Pauline Clance's work had described the experience as 'a new phenomenon and a unitary personality syndrome' (seemingly temporarily forgetting Kafka); Kolligian and Sternberg's research suggests that self-perceptions of fraudulence are 'a blend of inauthentic and self-deprecatory forms of thinking, with concomitant experiences of attention to one's behaviors and apprehension in evaluative situations' (Kolligian and Sternberg 1991: 323).

There are different models to explain how it develops. The one favoured by Kolligian and Sternberg is that individuals with perceptions of fraudulence are highly self-critical of themselves and are anxious about the evaluations of others. They feel a strong pressure to achieve and to excel. They are concerned that others will perceive their weaknesses, just as they themselves have. To reduce the possibility of exposure they closely monitor their behaviour and other people's evaluations and responses. They are constantly looking in on themselves – they know their weaknesses and how they try to disguise them. If they stopped monitoring themselves so carefully, they assume that others would see not just their failings but their multiple cover-ups, their fraudulence, their doubts and their attempts to remedy them.

Kolligian and Sternberg christened this as 'impostor syndrome', not the 'impostor phenomenon'. This might well have been what Kafka was suffering from. It is a set of processes based on an overly critical awareness of one's own weaknesses and shortcomings and the feeling that if you stop monitoring these so closely, they will leak out and everyone will see them, resulting in public humiliation. They are based around processes of attention, control and fear. Many psychological studies look only at specific aspects of this complex interplay, when it might well be that the connections between them are the most significant. That at least is a step forward.

- Feeling like a fraud is apparently surprisingly common in certain walks of life, particularly the arts, science and academe.
- We call this the 'impostor phenomenon' (or the 'impostor syndrome').
- In modern terminology, Franz Kafka suffered from 'impostor syndrome'.

- Pauline Clance and Suzanne Imes in 1978 first identified and labelled it in a sample of over 150 highly successful *women* – either 'respected professionals in their fields' or 'students recognized for their academic excellence'.
- Pauline Clance herself suffered from it.
- These women, the authors explain, were on every objective index high achievers but were unable to internalise their success and take credit for their achievements.
- The women would be concerned about new tasks and challenges because each new task could potentially expose them as a fraud.
- Women who suffer from it have a distorted self-image and cannot internalise success.
- They have difficulty accepting praise (because they think it isn't deserved) and are frightened of others discovering their lack of knowledge and ability.
- They often prefer low-level or less challenging positions so as not to be exposed.
- They also have ways of dismissing accumulating objective evidence of their underlying ability.
- A number engage in ritualistic behaviour to ensure success – superstitious rituals that can be controlled and can be used to change the focus of their attention as they seek chance but reassuring contingencies.
- The impostor syndrome is a set of processes based on an overly critical awareness of one's own weaknesses and shortcomings and the feeling that if you stop monitoring these so closely, they will leak out and everyone will see them, resulting in public humiliation.
- These processes are based around processes of attention, control and fear.
- Some have suggested an overlooked benefit of the condition, namely that this experience can drive one to try harder to succeed.
- Other researchers have reported the opposite – a correlation between impostor syndrome and self-handicapping tendencies, which involve placing an obstacle in the path of an evaluation so that possible failure can be blamed on the obstacle, such as not preparing adequately for an exam, rather than on yourself.
- That way you can protect your self-image because any failure does not reflect on you and your ability.
- You can self-destruct with impostor syndrome.
- The evidence strongly suggests that both men and women suffer from impostor syndrome.
- Indeed, impostor syndrome is found across all genders, races, social classes and ages.

- We do, however, need to learn a lot more about how and why it develops.
- Kafka had his theory about the role of the father, but there are several other accounts, all requiring further research.
- There is some evidence of a role for overprotective fathers with a narcissistic involvement in the child's achievements in the development of impostor syndrome (hardly reminiscent of Kafka's situation).
- From this theoretical position, the feeling of being an impostor derives from self-criticism, which is an internalisation of the parental desire for success.
- There is a lot still to learn about the phenomenon.

5 'I, the King'

I was in New York City, trying to make sense of somebody. Somebody famous. He clearly had a doubt issue but only in the sense that he never seemed to suffer from it. I was very near Trump Tower, but it was not the former president I wanted to see, even though he doesn't seem to suffer much from doubt either. I wanted to understand someone a little more (shall I say) complex and influential, an artist, a painter who has always fascinated me for good psychological reasons – not just because of his enormous and extraordinarily creative body of work (or the fact that he was a workaholic) but because of this apparent absence of doubt. His work was transformational and, according to many of his distinguished biographers, his personality and ways of thinking were critical to this, including or especially his lack of doubt.

He is at the opposite extreme to mine when it comes to doubt and self-doubt. I want to understand how that can be, where this apparent absence of doubt comes from and how it is maintained. If doubt is part of the conscious narrative of life, and often seemingly traceable to specific situations and specific events, what gives rise to absence of doubt?

This individual never had any doubt about his talent, his greatness or what he would achieve in life, even as a child, and these beliefs were almost certainly critical in allowing him to reach his potential. Famously, before he travelled to Paris for his first visit at the age of nineteen, he completed a self-portrait and signed it – not once but three times – 'I, the King; I, the King; I, the King.' As a child, it seems, he was indeed the king of a small, enclosed world of adoring women and these early experiences, many argue, were critical to the development of his self-esteem. Shadowy doubt, it seems, was kept at bay.

This painter was born in Malaga in 1881 and died in 1973. I say that he was born in 1881, rather in that year a child was stillborn. The baby's uncle, Dr Salvador Ruiz, leaned over the lifeless infant, exhaled cigar smoke into his nostrils, making the child stir, and that genius of painting, Pablo Picasso, burst into life. This is part of the great myth of Picasso. Narratives like this are important to how we deal with our lives, and Picasso himself liked this story. It was as if he was coming onto a great stage through the smoke in that room. Don Salvador himself also liked to tell how he breathed life into genius that burst fully formed from the womb. This story has become part of the narrative

DOI: 10.4324/9781003282051-5

of genius that every biographer repeats. Genius and great talent like this do not sit comfortably with doubt.

This child, the son of Jose Ruiz and Maria Picasso Lopez, was christened Pablo Ruiz. His father was an artist from a wealthy family and an art instructor in Malaga. By all accounts, he was a traditional and talented 'academic' artist, who believed in the discipline required for painting, and that art was a *craft*, learned through dedication, experience and practice. His paintings were meticulous, lifelike, the product of many years of study. Pigeons were one of his favourite subjects – pigeons in a loft, pigeons drinking from a bowl, pigeons front, back and sideways on, the feathers perfectly captured. Feathers are very difficult to paint, they require enormous skill and care, years of practice. He did these paintings so well that he was known by some in Malaga as the 'pigeon artist'. But the child was a different sort of painter with a different, precocious sort of talent.

His artistic gift was identified very early. Towards the end of his life, Picasso chatting with the photographer Brassai had this to say about these early drawings: 'My first drawings could never be exhibited in an exposition of children's drawings. The awkwardness and naiveté of childhood were almost absent from them. . . . Their precision, their exactitude, frightens me.'

The family always said that Picasso's first word was 'piz' for 'lapiz', the word for pencil, and that from the very beginning his lines and drawings were astonishing especially for a child not yet able to talk. He had been born into a household of women, his parents lived with his wife's widowed mother, her two spinster sisters and a maid – this was Pablo's first royal entourage. His mother had this to say about her child: 'He was an angel and a devil in beauty. No one could cease looking at him.' She told him that he was bound for greatness in any field that he set his mind to. 'If you become a soldier, you'll be a general. If you become a priest, you'll end up as Pope!' (reprinted in Mailer 1997).

But school was a different matter; there he struggled. He was terrified of being dragged off to school and surrounded by children of his own age which he couldn't command in the way he could those adults at home. He had great difficulty in reading and writing and didn't understand arithmetic, suffering from both dyslexia and dyscalculia (a maths learning disability that impairs an individual's ability to learn number-related concepts and perform basic maths skills). Arithmetic was such an issue for him because it was the form of the numbers, their shape that seized his attention – not the concepts behind the shapes. The number 7 to him was a nose drawn upside down, not a concept one up from six, and one down from eight. To him, the number 0 was the little eye of a pigeon; 2s were the wings, or 2 could be a man or woman kneeling in prayer. As a child, he lived in a world of form rather than abstract mathematical meaning.

He ran away from school whenever he could and hid in dark spaces in the home, escaping the terror. He liked to come out and sit surrounded by his adoring female entourage – drawing shapes to please them, especially spirals. Every time he drew a spiral he'd be rewarded by his mother or his aunts with a *churro*, which is a spiral-shaped potato chip. This, it seems, was part of his early learning – talent is rewarded, talent begets attention, like begets like – a drawing of a spiral leads to a spiral treat. It's the learning of a cause-and-effect relationship, a contingency, the shape of the art and the shape of the reward are similar – in primitive thinking, they call this the 'law of similarity': things that look the same go together. His father eventually arranged for him to transfer to a private school run by a friend of his, where he could draw as much as he wanted.

It's interesting to examine whether doubt did play any role in Picasso's life or whether it was entirely absent as many famous biographers have assumed. They say that his personality – self-assured, confident, without fear or doubt – was as important as his genius. So that's why I'm standing on 7th Avenue in New York City in the middle of summer, waiting for a cab. This was another step of my journey into doubt; I needed to see something, which may hold a clue about the great painter. I had made up my mind eventually to go there. As always, I had my own personal uncertainties but eventually decided that I really had to see this thing for myself. It might tell me something.

It was very hot and humid on the street, and I was sweating. Every time I looked up, I felt slightly dizzy. It was noisy and brash: horns hooting; engines roaring; loud, confident voices. Welcome to the Big Apple.

I was not forceful enough, not confident enough, I self-consciously waved hesitantly at the yellow cabs, and one by one they drove straight past to my less hesitant neighbours – their arms upright and resolute with no ambiguity. I already had to go back to my hotel once to change my shirt because of the heat. I was now running late. I had to get somewhere before it shut.

I had just given a talk on *risk* to a leading multinational but had avoided mentioning that voice in our heads, the 'd' word. Senior executives had jetted in from all over the globe to hear me. Not bad for someone who suffers from extreme doubt, I thought to myself, thank God I can hide it. The CEO of the company thought that his organisation, this household name with mul tiple products in all our homes, had become too risk-averse. He wanted me to explore the psychology of risk with his senior managers. 'Explain risk and risk-taking to them,' he requested. 'I want them to understand what's going on, what's holding them back.'

Oddly, with the scientific literature on this theme, doubt is rarely if ever mentioned. Doubt is a more intimate and, it seems, more private phenomenon; it's connected with risk and risk-taking, but it's broader and more insidious. Perhaps, when you are prone to doubt, everything is a risk, life is a risk.

So, I prepared the talk, without covering doubt – this most important of topics. I pointed out that risk-taking is central to human beings. As humans, we must take risks. It's central to human development through play and exploration, which are all based around taking small or large risks, and these are important for the development of language and cognition. Peek-a-boo teaches the infant about turn-taking, on which the regulation of turns in conversations are scaffolded. It involves the infant bearing the risk of going out of sight of the mother for a second or more hiding behind the flannel in the bath, or the mother covering her face with her hands. Watch the infant looking fearful as the mother hides for the first time and the joy when she reappears. The child learns quickly that this is a predictable sequence, of uncertainty and fear then resolution, a game. Risk-taking is central to relationships and survival. How can you find a partner without taking the risk of losing face? It is central to economic development; 'entrepreneurs' are, by definition, people who bear risks; the term was first used in the context of bearing financial risks in military operations. We are, I explained, a risky species.

The audience all nodded at exactly the right point. The CEO stared down at them from the side of the stage. So why then, isn't it easy to get people to take more risks? That was the point of the talk. It was purely rhetorical, although someone did put their hand up, and then slightly embarrassed lowered it again slowly and deliberately.

It becomes easier to understand, I continued, when you remember that risk is 'the chance or possibility of suffering loss, injury, damage or failure'. 'Many people do not want to experience even remotely thinking about the chance of suffering loss, injury, damage or failure,' I said. 'This doesn't feel nice, even just contemplating it, and when you start contemplating it, you may show how you feel, you may look weak, and others may see it. Your competitors may see it.'

I looked at these corporate bigwigs straight in the eye: 'this may be so important because others – competitors – may detect a *weakness* in you.' This sounded accusatory, but it wasn't meant to me. Perhaps, the display of weakness to our peers, to our immediate social group, is a fundamental aspect of our evolutionary past in groups with hierarchical social structures. Something to be avoided – even more so in a highly competitive multinational company.

'You might detect a weakness in yourselves and be even more conservative in future,' I continued. 'Your contemplation of risk and failure may prime negative memories – they become more available. You become flooded with things you thought you had forgotten.'

I had clearly hit some nerve. You could see this in their eyes for a quarter of a second, no more, below that glossy façade of the international globe-trotting executive, before they managed to cover it up with the ubiquitous practised smile. Psychologists call these emotional displays 'micro-expressions', quick emotional signals that emerge from time to time. Many people are oblivious to these, they go unnoticed and unchecked; it is my job to spot them. I use them as a guide when I'm speaking. You learn more that way.

'When confronted with a risky decision,' I continued, 'you may signal your weakness to others even whilst thinking about it' – it sounded confrontational as if I was accusing them of some basic faults:

> your heart beats faster, your blood pressure increases, your mouth becomes dry, your breathing becomes much faster, blood surges around your body draining away from areas like the stomach into the muscles and giving you 'butterflies', your pupils dilate so the moment can be perceived more clearly, your reserves of glucose are released to prepare for muscular activity.

You are now ready for fight or flight, or just sitting there in front of me today reflecting on all these physiological changes. Feeling exposed.

They looked guilty, as if they'd been caught out.

So how do we get people like you (I gestured their way to make sure they got the point) to take riskier decisions? These hand gestures are usually unconsciously generated, that's why they're so effective. But if you can fake them consciously and deliberately on occasion, with a little thought, they're very powerful (Beattie 2003, 2016). They knew that *they*, and their risk-averse decision-making, were what I was here to correct.

'You need to recognise that risk taking involves both thinking and emotion,' I said, 'and the relationship between the two – but both systems are flawed. And you must remember that emotion always drives the thoughts. Emotion comes first. We are not such rational creatures after all.'

They nodded awkwardly.

'But we need to consider both. When it comes to thinking about whether to take a risk or not,' I continued, 'this involves some sort of judgment of probability of things going wrong.'

'How good are you at doing this?' It was another rhetorical question but received a lot of below the breath answers, mouthed openly. 'The answer is not very,' I continued,

> because people judge an event as likely or frequent if instances of it are easy to imagine or recall. This makes sense because frequently occurring events are easier to imagine or recall than infrequent events. But some events are just more imaginable than others because they are emotional or highly visual, or both, so people overestimate their probability.
>
> If you ask people to estimate the probability of them *dying* from certain things [there were a few micro-expressions in the audience], they overestimate the highly imaginable ones like motor vehicle accidents, pregnancy, childbirth and abortion, tornadoes, flood, fire and murder, and underestimate the less visual ones like stroke, tuberculosis, asthma and emphysema. We're just not good at this. If I asked you to do this, right now openly and publicly, I wonder what *you* would come up with.

I made a half gesture, a stress-timed gesture on the word 'you', but not pointing their way, more a chopping movement which indicated that I wasn't going to put them on the spot, at least not just now. They looked genuinely relieved that I wasn't calling them out on this. Thankful, even.

> These vivid images may come from the media, film or TV – just think of how you thought about the probability of a shark attack after watching *Jaws* for the first time, but they are also generated through our own experiences, particularly our negative experiences. What do you remember best in your everyday life? The positive feel-good moments or the really

negative moments, emotional and shocking? Great successes that you've been working for or unexpected and devastating failures? The really good news or the really bad news?

They were too frightened to move now, lest I pick on them.

And what do you remember about bad news and the failures? It's not just the event or the outcome that you remember, that would be bad enough, but all the details of the context in which you experienced it – where you were at the time, who you were with and what you were doing.

It was like that night in the car park of the Royal Victoria Hospital in Belfast when I was told that my father had died. I can even apparently remember where everyone was standing after all these years, the rain on the tarmac, the words, the facial expressions.

All of the vivid details of these surprising and emotionally arousing events are recalled years, often decades after the event.

I explained that these were called 'flashbulb memories' by Roger Brown and James Kulik in 1977, a different type of memory from normal memory, which as we all know is faulty and forgetful. They argued that with flashbulb memories much of the vivid detail is captured for all time (although other research suggests that as in all memories construction and reconstruction plays an important role; see Conway 1994). These memories involve two of the most primitive parts of the human brain firing together – the reticular formation and the limbic system, responding to surprise and consequentiality. The argument in psychology is that these sorts of memories are critical for survival – their function is to remember the context of the event to avoid it in future. If you've been in a road accident or ever been seriously assaulted, you will have a flashbulb memory of the event, but we have them for other sorts of events as well, events that don't threaten our personal survival – the death of a loved one, the death of a public figure (Princess Diana; JFK, if you're old enough; Michael Jackson), and other things that haven't worked out, shock-ing and surprising and important – failing an exam, a bad deal, a terrible row. Just recall where you were and what you were doing. Most people have vivid flashbulb memories of these.

I continued:

If you have a flashbulb memory for a risky decision that goes horribly wrong and unsurprisingly wrong and you have a strong emotional reaction to it, then you will have a very vivid flashbulb memory of the event. If something is so vivid then you think that it's recent and common and then you will overestimate how likely it is and the probability of it happening again. That's why you won't then make risky decisions.

This is the joy of lecturing. You're in control for a while, you're implanting ideas into people and watching the machinations of their minds in their chang-ing facial expressions. And you stand there, behind a lectern, covering up.

None of us is immune to these detailed and vivid flashbulb memories (even when slight errors slip in), although we might not always like to admit it, and the fact that they are triggered by sometimes seemingly trivial negative events does not help. I have a flashbulb memory of a coach saying several months ago that I looked slightly overweight (I'm a keen runner and he was just trying to help) – not just the words ('you've got to drop the baby') but the situation (where he was standing, what he was wearing, where my fellow athlete was standing) are encoded in my brain. I remember little else about the session. We do need to try to understand how certain events (including seemingly trivial events) become sufficiently consequential, surprising and emotionally arousing to trigger these vivid and enduring flashbulb memories memory in the first place. (It might be relevant that my beautiful cousin Myrna died of anorexia just after my father's death – it was said that her illness was triggered by a chance remark about her weight by a doctor. She is buried in the next row to my father in Roselawn Cemetery.)

But, it seems, we are all constrained by vivid memories with implications for how we think, reason and behave.

It was odd lecturing in New York to these successful businesspeople, talking about the limiting effects of flashbulb memories, and yet being so acutely aware of my own personal doubt-ridden mental life, and yet hiding it to the best of my ability.

Of course, I explained what they should do to feel more comfortable taking risks – they should get happy, as I explained in my book *Get the Edge* (Beattie 2011). Depressed people are more conservative in their choices than non-depressed people. Students who have just read about a negative event showed an increase in their estimate of all kinds of hazards. Happy induced students showed a decrease.

> Make yourself happy by consciously recalling the five best things that happened that day as you brush your teeth at night – prime positive memories – wake up in the morning happier. And when risks do go wrong after something you've done, don't internalise the failure, don't think that it's always going to be present and affect everything you do, and that it's all down to you. Change what is called your 'attributional focus'. Don't think about setbacks and failure in terms of 'always' and 'never'. Think about them in terms of 'sometimes' and 'lately'. Use qualifiers and blame bad events on ephemera. To take more risks, you need to re-adjust cognitive focus, and focus more on what may be achieved instead of what may be lost. Recognise that you have a hard-wired emotional system that is going to affect you and your decision-making. Learn to enjoy the physiological feelings of nervousness as excitement.

'And what about flashbulb memories for failure?' I continued:

> That may be harder. You need to prevent the formation of flashbulb memories because these will always cognitively bias you to avoid risk. To

avoid such flashbulb memories, you need to reduce the surprise, emotional valence and the consequentiality. But how can you do that? Would not whether you internalise the failure be critical here? Is that not what makes it so consequential? That surely would affect the 'consequentiality'. If you blame yourself for things that go wrong, then when they do go wrong, they will have a much bigger effect on you. One aspect of that will be the formation of flashbulb memories.

I left it at that. 'Finally,' I said. My take-home message was remarkably simple and clear. The kind of message that is found in every self-help book; in fact, it *is* found in every self-help book.

'Treat negative thoughts as if they were uttered by another person – a rival whose goal it is to make you miserable. Argue back, tell them to get lost. Only then,' I said, 'will you be willing to embrace risk.'

There was a lot of applause, which was just as well because the voice in my head was telling me that this was all a little too 'schematic'. The voice in my head wouldn't get lost, no matter how forcefully I told it.

So, what was I saying? When it comes to risk, both thoughts and feelings are important and both processes are subject to bias. Your memories, your personal memories, are critical – they will influence your behaviour. But highly personal memories are rarely probed in contemporary psychology; it relies instead on generalisations here and in so many other places. Most flashbulb memory research is based on big news events common to whole sections of the population (e.g. the death of major public figures or events like 9/11). But what about mundane flashbulb memories triggered by slight insults ('drop the baby') which can only perhaps be understood in the context of a life.

And what about making yourself happy by consciously recalling the five best things that happened that day – prime positive memories. A mantra for our times; you see it everywhere, in every huckster psychology guidebook. You always must do it when you're brushing your teeth at night. But what about individual differences in all of this? Does it work for everyone? How representative a sample of ordinary people leading their ordinary lives would you need to even test this? Has it been tested on busy executives like those in front of me, suffering from the inevitable jet lag given their lifestyles of continuous international travel?

And as for that voice in your head and treating it like it was being said by another person, sometimes, maybe often, the voice in my head is my friend, trying to help. That voice in my head is sometimes my doubts when they come into consciousness, emerging from that vague emotional feeling which sometimes cloaks them. Psychology has uncovered some of the processes connected with risk-taking, which presumably have *some* implications for the origins of doubt, but it seems to have much less to say about why we all vary so much. It's a broad, schematic overview of the processes. The details in individual lives must sometimes be critical.

So that's why I was waiting for the cab on that New York street. I was still sweating. I had forgotten my deodorant and there was a wet patch under my arm, the one that I was holding up – I checked, under both arms. A cab stopped. I asked the driver to take me to the Museum of Modern Art on West 53rd Street. The car felt cold, the air conditioning was up full blast, the cab smelt of lemon air freshener, it was too strong, but I was even more embarrassed about my sweat stains, and the driver started chatting almost immediately, a bit too over-friendly for a man like me.

'Whatcha going to see at MoMa?' he asked. It took me a few seconds to work out what he was talking about. I was thinking about other things, nervously thinking that when I got to the Museum of Modern Art, I might learn nothing.

'Picasso,' I said. I hate direct questions like his about why I'm visiting a museum or gallery. Any response sounds pretentious. I'm very sensitive to what others think.

'Great, buddy,' the tone was exuberant, overly exuberant for me. 'Anything in particular?'

'I'm interested in *Les Demoiselles d'Avignon*; that's what I'm hoping to see,' I sounded both hesitant and pompous at the same time. My mother always said that it was hard to tell that I was professor of psychology with a doctorate from Cambridge. She said that I never sounded like one or looked the part either. I tried to enunciate '*demoiselles*' correctly, but it didn't sound right, it didn't sound French enough with my working-class Belfast accent and adding 'that's what I'm hoping to see' made it sound worse, like I was chattering, filling time. I wanted to sound like an academic in control, but the words came out making me sound like a pompous fool. The phony French accent with a Belfast twang made it worse.

'You're heading to the right place, buddy,' said the driver, even more exuberantly. There was a brief pause. 'I've heard of that painting, you know. I picked up a fella talking about that just the other day. He was arguing with his wife about it. That's how famous it is. That's the painting of the prostitutes with the masks on, right?' he said, but even with the 'right' at the end, it didn't feel like a question.

'That's the one,' I said.

'So, what do you do then, buddy?' he asked. 'Are you into art?'

'Well, I'm a psychologist,' I replied, including the 'well' to make me sound both modest and broader in my interests, not defined by my position.

'A psychologist, oh,' I could see him making faces in his rear-view mirror, as if to say I get all sorts in my cab. Psychologist is no big deal, yesterday I had a criminologist and the day before a taxidermist. The 'oh' was short, it suggested no big deal.

'Maybe you can help me then. Why are they wearing masks?' the driver said. 'That's what they were arguing about. He had one theory, his wife had another, she asked me to be the judge.'

I laughed politely. 'Whose side did you come down on?' I asked.

I'm a New York cabbie, not some highbrow art connoisseur, for Christ's sake, I just humoured them. I googled the painting when I got home to have a look at it. I like to pick up bits of knowledge. But I thought to myself – that's some messed up piece of shit to be honest. I'm not sure I want to pay good money to go and look at that. Sorry to spoil your day, buddy. There are better things in MoMa than that. His other stuff has gotta be better than that.

I noticed that 'gotta' was emphasised and extended, as if he was trying to show me some sympathy and some direction.

I didn't respond.

'That Picasso guy is a genius, right?' He added the tag 'right' again to force me to say something. It's meant to be what the powerless use, to elicit agreement, but in some contexts, it's quite the opposite. It's a form of control. That's what I was thinking but I still felt compelled to respond.

'Right,' I said, 'that's what they say.' For the moment, as if trying to exclude myself, lest my views come under direct scrutiny in this confined and now claustrophobic space.

I was starting to think that maybe I should study the driver who had no inhibitions and was not a fellow doubt sufferer. I could see that he was smiling enjoying our chat. He would leave the conversation feeling better about himself, with something to say about an anonymous fare with sweaty arm pits.

I had this fella in my cab, some kinda psychologist, going to see that Picasso painting, the one with the pros in the masks, he had no idea why he was going, I told him not to bother. He was sweating a lot; I should have told him to take a shower.

I was feeling a little worse. Maybe, I overthink situations.

I knew a little about Picasso's background. Some events struck me as potentially critical to the issue of doubt. I knew that his family had moved to Barcelona when Picasso was fourteen when his father was awarded a position in the La Llotja School of Fine Arts, making Pablo an outsider – an Andalusian in this sophisticated city in Catalonia, barely able to understand Catalan or make himself understood, despised by middle-class Catalans. A despised outsider. Pablo's long-term friend Sabartes reminds us that 'Andalusian [is] never pronounced without a grimace of repulsion' in Catalonia. Picasso's self-importance would develop into a strong view of himself as an outlaw, a fighter prepared for the artistic battles that lay ahead.

At fourteen, Pablo entered his painting *First Communion* for a local competition, and it was exhibited at the Exposicion de Bellas Artes with other famous painters from the region. This repulsive, swarthy Andalusian was already being recognised for his potential.

At twenty, he did something shocking in his social and cultural context. He decided to change his signature to something which was in his mind more suitable for a grand artist. Pablo Ruiz became P. Ruiz Picasso, then P.R. Picasso and finally Pablo Picasso. He explained this in conversation to Brassai in 1943:

'I'm sure [what attracted me to it] was the double "s" which is very rare in Spain. . . . Have you ever noticed that there's a double "s" in Matisse, in Poussin, in Rousseau?'

There may have been some slight argument for this in terms of the 'feel' and alliteration of the name, the name reengineered to become a symbol linking him to great artists, but its broader personal and symbolic significance couldn't have been lost on him either. He was reinventing himself, removing the bloodline of his father, deleting his father from his art and his being. Bloodline signalled through the father's surname was even more important in Spain at the turn of the twentieth century than it is today, but especially so in Andalusia. Through this one act he was rejecting the link with his father and choosing in preference his mother's name with all the hurt that would cause. It has been pointed out that physically Picasso resembled his mother much more than his father. His father was tall, thin and blue-eyed and often called 'the Englishman' by his neighbours; his mother, on the other hand, was short, dark and thickset like Pablo himself. Pablo was rejecting his father with the name change, and rejecting his father's art, and leaning out to touch the giants of the art world. He was not going to be someone who dabbles at art, or paints pigeons for small commissions.

This change in name must have caused considerable hurt to his father. How could it not? It has been suggested by psychologists that the animosity towards his father went deeper than this issue of mere nomenclature, some have evoked the Oedipus complex to explain it. In classic Freudian theory, the Oedipus complex is used to explain the intense competition that can arise between a father and son. Freud argued that it emerged from an unconscious desire by the child to kill the father to possess the mother. Biographers and art critics point to other actions by Picasso to support this idea. Picasso had painted over his father's unfinished sketch of a pigeon when he was thirteen; it almost led his father to give up painting. Picasso obliterated his father's art; he removed the public display of his father's biological legacy by changing his name. But there were other things as well. According to a long-term friend Tristan Tzara, who was a friend of Picasso for over thirty years, recalling some of Picasso's early experiences, Picasso had 'made himself a man' with a tall, thin waitress up against a barrel in the bar above which he had his secret studio. Picasso had later described losing his virginity in that bar as like 'like fucking his father'. It's hard not to see some Oedipal significance in all of this, or at least traces of some sort of unconscious drivers at work.

Coincidentally, it's important to note that in contemporary psychology, it seems that we can talk more freely again about the role of the unconscious now reformulated in certain areas of psychology in terms of systems of thinking by the Nobel laureate Daniel Kahneman. Kahneman's System 1 is automatic, quick and largely unconscious, and this will play a role throughout this journey of ours, especially when it clashes with the more rational and conscious, and slower, System 2 (Kahneman 2011; Beattie 2013, 2018a; Beattie and McGuire 2015, 2016, 2020). When we consider the concept of doubt in any detail, we will have to talk about conscious and unconscious processes because sometimes when it comes to doubt, we are consciously aware of this phenomenon, but at other times, it's just below the level of consciousness but still exerting its effects on behaviour – in terms of delay and procrastination, and then only slowly emerging. Whether we subscribe to particular theories of the unconscious, like Freud's and his concept of the Oedipal complex, is quite a separate issue, but there clearly was something of psychological interest in Picasso's attitude to his father.

But this change in signature and identity by Picasso, many biographers conclude, is also particularly important when it comes to the issue of doubt. It is reminiscent of the boxer Cassius Clay years later proclaiming that he was 'the greatest': it is a public display of bravado, a public challenge – you are setting yourself up for others to attempt to knock you down. Picasso the self-styled Matisse, Poussin, Rousseau; he was to be the greatest. You cannot have doubts to do this. In Picasso's case, it was even more extreme because in changing his name from Ruiz to Picasso, he was symbolically and publicly cutting himself free from his father. This was to be the Picasso that the world would come to know, no longer the son of Ruiz – born again, a man without a clear link to his father, a genius out on his own.

There would *seem* to be a pattern in all of this. The early experiences of adulation in the home by the females of the house, the rejection at school, the outsider in Barcelona, the fight back – the reliance on one's own talents and vision, the lack of doubt and confidence in cutting bonds and signalling a new identity, the clarity, the recognition, the emergence of genius. Doubt in this narrative might seem to be the enemy of genius, the great inhibitor; genius needs confidence and absence of doubt to develop and flourish.

But there are other things about Picasso's life that sit uneasily with this narrative. Other small events that are often interpreted in a particular way to fit in with this overarching narrative. When Picasso returned to Barcelona from Paris, poor and depressed, he began to paint the great works of what was to become known as the Blue Period. The first few months back were very difficult, with Picasso now suffering from a major depression. More than forty years after the event, one of Picasso's last lovers, Geneviève Laporte, described his return to the family:

> I learned one day how Picasso, exhausted by the poverty of his life in Paris, had returned to his parents' house in Spain. He arrived in the evening and worn out by the long journey, went straight to bed. The following morning, while he was still asleep, his mother brushed his clothes and polished his shoes so that on waking Pablo found that he had been deprived of 'his dust of Paris'. He told me that this put him in such a terrible rage that he almost reduced his mother to tears.
>
> (Laporte 1975: 19)

This is often understood to be a story of the fury and impatience of genius, Picasso fighting his way out of his depression to realise his genius, but there is another way of thinking about this. This is superstitious behaviour, one of the earliest forms of psychology, an attempt to understand and predict the machinations of the world, to keep anxiety and uncertainty at bay. More specifically, it is a form of what the psychologists Paul Rozin and Carol Nemeroff have called 'sympathetic magical thinking' – universal and primitive forms of thinking diametrically opposed to current scientific beliefs, but with their own 'laws' underpinning them. Edwin Taylor in 1879 outlined some of the laws on which sympathetic magical thinking is based. Firstly, there is the 'law of contagion' which holds that 'Once in contact, always in contact' – when people or objects make physical contact, the *essences* of them may be permanently transferred (Rozin and Nemeroff 2002). The essence of Paris was in that dust and on his shoes, which were now part of him. Cleaning those shoes could take that, and the creative influence of Paris, away from him.

Parts of the body also contain the essence of a person. In the words of Rozin, 'fingernail parings contain the "essence" of the person to whom they were previously attached, and foods carry the "essence" of those who prepared them.' According to Picasso's friend Sabartes, we learn that Picasso could not bear to part with his fingernail clippings and in 1903, he wrote the following inscription, 'The hairs of my head are gods like myself, although separated from me.'

This magical thinking underpinned his life. This is how he dealt with doubt, by taking control though superstitious behaviour – by allowing superstition to govern much of his everyday behaviour.

Critics point out that his early work was highly derivative in terms of style, and it was not until his friend Carles Casagemas' suicide in Paris because of a failed relationship that Picasso produced what is considered by many his first major original work – a painting of his dead friend in his coffin, a flickering candle behind him but prominent in scale, the whole painting a simple oil on panel measuring just eleven by fourteen inches. The flickering candle looks to many like a vagina, perhaps signifying the vagina, the lover, that had led to the suicide, but the candle also signifies the sorrow of Casagemas' mother in her lonely, grief-stricken vigil. The candle as a signifier of the Catholic Church and carnal knowledge in one. This one image laden with symbolism.

This is recognised as the first of Picasso's paintings to lead us into a deeper way of thinking about the scene in front of us, the precursor of the Blue Period and then on to the vast aesthetic range of Cubism. The unconscious has been given free rein.

He follows the first painting of Casagemas with a second painted just using blue, and then another portrait in blue of his friend Sabartes. This was the birth of the Blue Period. Picasso's subsequent long and deep depression throughout the period was occasioned quite possibly by his personal guilt for his friend's death – he himself had encouraged Casagemas to leave Malaga to join him in Paris. He had been partly responsible for the move, and then after his friend's death Picasso had a relationship almost immediately with his friend's lover.

People react to trauma in different ways. Depression is common, and Picasso's depression was so severe that for several years he was close to suicide on several occasions. Doubt and inaction arising from a depressed state are not uncommon, but that was not Picasso's response. This sense of his own greatness, propelled no doubt by that early adulation in the household and early recognition of his unique talent, allowed him to express his feelings through his art in the midst of trauma – to create a unique art form. He managed any emerging doubts about life, death, love, in his own way. But he became highly superstitious throughout. Superstition can be construed as this most primitive form of psychology, an attempt to take control of your destiny when control is wrested away from you. You become conscious of small things that may have a much greater significance than might at first appear and look out for possible chance occurrences and co-occurrences. You notice the black cat, you look out for some evidence of good luck, and you bind them together in a neat superstitious bundle. You may repeat actions, just in case you haven't got it right the first time, to make sure that the action perfectly executed have the right effect. You become emotionally dependent on small inconsequential things.

Picasso flew into that rage when he returned home to Barcelona from Paris after his mother polished his shoes because she had deprived him of 'his dust of Paris'; he needed that dust of Paris on his shoes to fire his imagination. Anything that came into contact with his body or any bodily part, hair, fingernail clippings, became important to him, significant, sacred almost. That is one aspect of magical thinking.

In 1903, whilst painting the white walls of his friend Sabartes' apartment on the Carrer del Consulado, he had written, as we have seen, 'The hairs of my head are gods like myself, although separated from me.' It was neither a joke nor a provocation; it was a proclamation. Here was an artist who despised doubt, God-like in his power. It was a proclamation of his greatness, but to whom? To Sabartes? To himself? Was this indeed an example of what psychoanalysts would call a *reaction formation*, a chance to deal with any underlying anxiety about status and ability and greatness, any self-doubt, by fixating in consciousness an idea that was the exact opposite of the feared unconscious impulse? And was that the basis of him signing the painting 'I, the King' three times, and changing his name until it superstitiously mirrored the names of Matisse, Poussin and Rousseau, with its double letter 's'? And was that the basis for him clinging on to the dust of Paris on his shoes in case his greatness could be blown away by the wind? What doubts lay beneath the surface that may help explain his extraordinary *productivity* of this icon over so many decades?

Another law of magical thinking is the 'law of similarity', which holds that 'like causes like', or that causes and effects should look similar. For many primitive peoples the cause of the disease should have a surface relationship, a similarity, to the symptoms. Thus, the Zande people of Africa believe that fowl excrement causes ringworm because fowl excrement and ringworm look similar. The Voodoo practice of burning a representation of an enemy to cause them harm is another example of the law of similarity. The doll looks like the person, and thus the person can be harmed by burning it. Among the Bantu people there is a belief in *kuntu*, the spirit present in both people and things, and the way that they combine and intermingle. African masks have *kuntu*, and when they are worn that spirit transfers to the wearer.

Long before the appearance of African masks on the two women to the right of the painting in *Les Demoiselles d'Avignon*, Picasso had become preoccupied with African masks and their power, ever since he came across them in the Ethnological Museum in the Palais du Trocadéro. This was never simply about copying form, as some have assumed. The painting itself has five figures compressed onto the canvas, all with sharp jagged edges, like shattered fragments of glass. The one on the extreme left resembles pre-classical Iberian sculpture (Picasso had recently bought two archaic Iberian heads stolen from the Louvre, so he had something to copy from); the two on the right wear the masks.

In their essay on the painting, the art historians Beth Harris and Steven Zucker write: 'Because the canvas is roughly handled, it is often thought to be a spontaneous creation, conceived directly. This is not the case. It was preceded by nearly one hundred sketches.' In earlier versions, there were two male figures in the painting, a sailor in uniform in the middle and on the left a medical student in a brown suit, carrying a textbook. It has been suggested that these two figures missing from the final painting symbolised two different sides of Picasso. The sailor, at sea for months, represents the lustful Picasso – in the original sketches the bowl of fruit between his legs is lengthened and sharpened, indeed phallus-shaped, and pointing to a particular prostitute in the middle, the object of Picasso's desire. The medical student symbolises the more analytic Picasso, analysing how their bodies of the prostitutes are constructed and assembled.

But there is a different interpretation, favoured by William Rubin, the senior curator of the Department of Painting and Sculpture at the Museum of Modern Art. Amongst others, the medical student recognises disease; the masked figures look alien, unknown, aggressive; they instil fear, African masks on white bodies. The painting for some is about Picasso's desire but an expression of his fear. He is terrified of the diseases that they would transmit to him. We know that Picasso visited brothels in this street, Avignon, in Barcelona; he understood the risks of contracting syphilis. Those figures on the right of the painting are not wearing masks as adornment – that is their character; that is *kuntu*. Indeed, decades later he would tell Andrew Malraux:

> The masks weren't just like any other pieces of sculpture. Not at all. They were magic things. . . . The Negro pieces were *intercesseurs*, mediators; ever since then I've known the word in French. They were against everything – against unknown, threatening spirits. . . . I understood. I too am against everything. I too believe that everything is unknown, that everything is an enemy! Everything! Not the details – women, children, babies, tobacco, playing – but the whole of it! I understood what the Negroes used their sculpture for. . . . They were weapons. To help people avoid coming under the influence of spirits again, to help them become independent. They're

tools. If we give spirits a form, we become independent. . . . I understood why I was a painter.

(Malraux 1974: 18)

Magical thinking may not only hold the key to understanding aspects of the painting but also highlight some of the primitive superstitious processes necessary for Picasso to keep doubt at bay, to allow him some degree of control in a changeable and hostile world. In his outburst with his mother when these processes were interfered with, we may see through that façade of extreme confidence. Perhaps this is a hint that the complete absence of doubt and self-doubt might sometimes just be on the surface, requiring complex structures and beliefs to control it. These also need to be explored, and like doubt itself are normally hidden away (great art allows them to be articulated).

Coincidentally, these laws of magical thinking are not restricted to primitive peoples and great artists. Rozin's research also tells us that we can see evidence of this even in the most 'rational' of groups. Most university undergraduates, for example, abide by the laws of similarity when they are confronted with powerful enough emotions like disgust (a very powerful emotion because of its evolutionary significance). University students studied by Rozin et al. (1986) would not put fake vomit in their mouth that was clearly made of rubber. It was what it looked like that dictated their response; logic could not override this. They also would not consume sugar from a bottle labelled 'Sodium Cyanide, Poison', even though they themselves had arbitrarily put that label on the bottle. The students wouldn't even consume the sugar when the label read 'Not Sodium Cyanide', which supports Freud's idea that the unconscious does not process negatives.

My friend Duck back in my gang in Belfast understood the power of the laws of magical thinking, and the law of similarity. He was the joker of the group, and when his mother was baking (which was a very rare and very special occasion in any of our houses) and when she wasn't looking, he liked to make a chocolate cake in the shape of a dog turd. He brought it down to the street corner wrapped in tin foil, careful not to damage it in any way. He unwrapped it with a great theatrical gesture and sniffed it. He made an 'mmmmm' sound. It smelt of chocolate, but nobody would eat it even when he tried to ram it down people's throats.

'It's not dog shite, just eat it.' I can see him now, the chocolate melting in his hand, like runny shite. Everybody ran away, screaming like children.

The funny thing was that he couldn't even eat it himself. He tried but he couldn't swallow it. He loved chocolate but not like this.

Perhaps, we all have these laws of magical thinking in our survival repertoire overwritten by science and good sense. They centre on major emotions like 'disgust', which originally kept us alive by getting us to avoid certain offensive foods, but then went through a period of cultural evolution to other aspects of our existence (sex, death), then on to certain types of moral behaviour. But disgust and other strong emotions still hinge on these very basic principles like similarity and contagion.

What we also know from Paul Rozin's research is that there are very large individual differences in the extent of sympathetic magical thinking both within and between cultures. It may give us solace and protect us from doubt. Perhaps that was part of Picasso's secret.

If we really want to understand how doubt sits within the individual, we need to begin to probe and uncover these mechanisms, as shadowy as doubt itself. Some people may rely on science and logic to offset doubt, some might use more primitive processes that they barely understand. That's why in trying to come to terms with doubt we may need to consider how the brain works in terms of systems of thinking, both rational and non-rational, explicit and implicit. We might also need also to consider the notion that the two hemispheres of the brain operate in very different ways, and that some primitive processes of which we know little are hard-wired into one hemisphere.

So why did I want to see *Les Demoiselles d'Avignon*? Because I think that it may have doubt written all over it. It may be hailed as a work of genius and the first great work of Cubism, but it feels to many unfinished, as if Picasso had doubts about what he was trying to create. And we need to remember that Picasso wouldn't allow the painting to be exhibited for nine years after it was painted. I wanted to see this doubt for myself. I wanted to be reminded that doubt may not necessarily be in the explicit verbal narrative of people's lives, something that they choose to report, but may be visible in the action narrative of their lives. I may be something that can only be ascertained when we consider and analyse their actions. As Kahneman points out, many of our actions are not guided by conscious processes. If we are to understand doubt, we may have to consider some non-conscious processes and how they manifest in terms of behaviour.

The only problem was that the museum was closed by the time I got there. I had prevaricated too long on my decision and not been direct enough in getting that cab.

Doubt, you see, was ruining my life. I, for one, could not keep it in check.

- Many great innovative and transformational artists, like Picasso, who took great risks in their work are said to be extremely confident and not to suffer from doubt. I had to examine this idea carefully.
- Human beings are a risk-taking species, and yet doubt inhibits many.
- We need to take risks for our cognitive development, in order to develop relationships or for economic success, but many find this difficult.
- Risk-taking is affected by both emotion and thinking; emotion comes first. This may be very important.
- We have detailed imagistic memories which may affect our judgments about the probability of things going wrong.

- These may lead to severe doubts.
- If you are happier, you take more risks.
- It is said that is you can prime positive memories last thing at night by remembering the best things that have happened to you that day – that way, it is claimed, you may wake up happier.
- This would mean that you should be more comfortable in taking risks and having fewer doubts.
- If you don't internalise failure, that is avoid assuming that failure is all down to you, then you should be able to take more risks.
- There is wide individual variation in all of this.
- It turns out that Picasso was extremely superstitious and used this to keep doubt at bay.
- If we want to understand doubt, we need to understand the mechanisms that people use to cope with it.
- Some of these may be rational; some may be highly irrational.

6   The perils of a doubt-free life

The house is at the top of a hill. I found it quite easily. It is surprisingly grand, not like the house of the imagination or the films about one of its previous occupants. Toys spill out into the garden. Bright red and yellow, tractors and cars. It's a family home. Probably a few kids knocking about the place. It's a semi-detached house but very spacious. It's just less than a mile from the station. A good distance, too short for a taxi, just a bit too long for a walk, especially with that hill, and in the Manchester weather where it's always raining. Even though it's not in Manchester. It's a bit further away and a lot more private than a house in the Manchester suburbs.

The old fuddy-duddies from the university wouldn't be calling unexpectedly at this house, they wouldn't just be dropping in on the way home from that Gothic pile on Oxford Road. They would need to walk up that hill where the pavement suddenly stops. It would take some effort. It is very private or self-contained, the way a young family might like it, or someone who liked their own company, or their privacy. Someone who didn't like to be disturbed.

It would suit a runner, many people would probably miss that about the house. The river Bollin stretches in both directions at the bottom of the hill, 478 strides down the hill to be exact, of course I counted them, I am a runner after all, and you always must be exact when it comes to distance and time when you are one. When you go down to the bridge over the river, you can run either to the left or to the right. A mile and a half if you go left where the river meanders through some meadows, past some sheep and then some cows, a white bull in the spring sunshine amongst them, flat, but rough underneath, hard to get a good pace, but demanding for training. Or you can turn right, less flat, with small hillocks that you can sprint up, and the river Bollin bubbling away, a small waterfall that draws you to it with the sound in the distance. Another mile and a half before some large broad steps and only then some houses. It would be good for private training where you are not watched or scrutinised. You can go both ways, and do six miles, or twelve, by doing it twice, without seeing anyone some days. Running, thinking, counting out the distances, it's impossible to stop that counting, of course – distance, time, velocity, blocks, categories, results, data. Numbers underpin running, but then again, numbers underpin everything.

DOI: 10.4324/9781003282051-6

They had made fun of him at school for the amount of time he spent in his solitary cross-country running. 'It'll stop him wanking for a while, the arrogant little shit,' his fellow pupils said. Even the teachers commented on it, along oddly similar lines. 'At least he'll use up some of that . . . energy,' with that long polite pause before the word 'energy', and the word itself always enunciated in a certain way, and the teachers would laugh, as he set off past the chattering boys for another run, in solitude. He never seemed to feel the cold, as if running itself was a form of chastisement. He liked to run in singlets and baggy shorts, grey socks, floppy trainers, at school in winter and even later, when he was running from that house on the hill near the Bollin. As a student at Cambridge, he would go for long runs along the banks of the Cam sometimes as far as Ely. Running settles the mind; it helps it focus. It was sometimes in early summer in 1935 in Grantchester meadow after a long run that he lay in the sun and started to doze off, thinking about mathematical proof and thinking about trying to answer them using 'a mechanical process'. In that dozing, dreamy pause in the middle of that run, he was dreaming of machines, and this idea would eventually turn into the concept of the 'computer' in our modern sense. When Alan Turing originally used the word 'computer' in 1935, he – like everyone else – meant a *person* doing calculations, but soon he was suggesting the possibility of constructing a machine 'to do the work of this computer'. This itself was not that novel. Turing had read books as a child which talked about the brain as a machine, a telephone exchange (another very common metaphor when we still used them) or an office system. Babbage had designed the 'Analytical Engine', a universal machine for doing calculations a hundred years previously. But what was original about Turing's ideas was that he combined 'a naïve mechanistic picture of the mind with the precise logic of pure mathematics. His machines – soon to be called *Turing Machines* – offered a bridge between abstract symbols and the physical world' (Hodges 2014: 138). His idea was reported in his paper 'On Computable Numbers', published in 1936. He was disappointed by the reception to the paper – only two people asked for reprints of the article in the months that followed. Some have assumed that he had little doubt about the importance of his fundamental idea even if its brilliance was not immediately recognised. Others are not so sure. Surely if it was that important, others would want to see it immediately. Turing vacillated between both positions.

Sometimes people would stare at him as he ran – it happened at school, at university, and now here in Wilmslow. He was used to it by now. He paid them no heed. Running was a daily ritual; he had once broken his ankle when he had slipped on some wet stones, later he had a severe hip injury, all that wear and tear, but none of this stopped him. He had many rituals to get through life – running, always wearing a gas mask to keep hay fever at bay whilst riding his bicycle when working at Bletchley Park during his code-breaking days, always wine with dinner now he worked as a Reader at the Victoria University of Manchester ('I'm now a Reader and a writer', he would chuckle over dinner), the wine mulled in hot water every night, the cork always put in the bottle after the meal was finished, no matter what any guest might want. Rules come first before the individual taste or needs or wants of others. Rules help keep doubt and uncertainty at bay. It has been suggested that some people with similar personality characteristics as Alan (poor social skills, poor understanding of the mental state of others, poor reading of emotional expression, a focus on systems rather than on people, etc.) have a high intolerance of uncertainty. There are two main dimensions to this construct, each with somewhat different implications. The first has been called 'desire for predictability', which refers to a dislike of unexpected events and a need to make the future as certain as possible. The second is more extreme and sounds it – 'uncertainty paralysis', where people can feel cognitively and behaviourally *stuck* in the face of uncertainty (Boulter et al. 2014). Poor social skills and difficulties with understanding everyday social communication, of course, increases the uncertainties surrounding social situations. This makes the individual more anxious, which in turn inhibits their behaviour even more. Speech becomes hesitant, filled with voiced hesitations, more faltering, less sure. People commented on Turing's filled hesitations, his 'ums' and 'ahs', and the shrill tone of his voice, and also far too loud in the wrong social contexts – for example, in private conversation. His booming, shrill voice drawing attention to itself.

Of course, there is an extraordinary irony in all of this. Someone using rituals to keep uncertainty at bay and simultaneously working on the mathematics to allow a machine to compute like a human being, which will change our world forever, and in so doing, herald the most unpredictable of futures. Artificial intelligence – many still find this concept somewhat chilling; they say that we can't possibly know where it will all lead.

A close friend had once said to him that people cannot change. People are how they were born. It was a work colleague, Joan Clarke, at Bletchley Park, that top-secret establishment – she had made the remark during the Second World War. They had been engaged once, but only for a very short time. It was an odd sort of engagement. The day after he asked to marry her, he told her that he needed to explain something to her. In the words of Alan Turing's nephew Dermot Turing:

> For the first time in his life, Alan Turing found in Joan a woman he could talk to, on the same level, despite her sex. They tended to do the late shift together; and after some months of chess-playing and general socialising, Alan Turing turned the unthinkable into the unimaginable.
>
> (Turing 2016: 117)

As Joan Murray (nee Clarke) said in an interview in 1991 for BBC TV's *Horizon*, 'The Strange Life and Death of Dr Turing':

> I suppose the fact that I was a woman made me different. We did do some-things together, perhaps went to the cinema and so on, but certainly it was a surprise to me when he said – I think his words probably were "Would you consider marrying me?" But although it was a surprise, I really didn't

hesitate in saying yes. And then he knelt by my chair and kissed me. We
didn't have very much physical contact.

But the next day after lunch, they went for a walk together and he said that
he had to get something off his chest. He told her that he was a homosexual,
but he had tried to soften it a little. In Joan's words – 'he told me that he had
this homosexual tendency, and naturally that worried me a bit, because I did
know that that was something that was almost certainly permanent. But we carried on' (BBC Horizon: 1991). She called the engagement off eventually – she
wasn't angry, a little surprised, but not angry. It's in your nature, she had said,
you cannot change. You don't have a 'tendency' towards homosexuality. And
life then followed its natural solitary course.

But so too was running – that was in his nature as well, and the house allowed
him to run still, after what he had been through – to run still in in private. Of
course, he knew that his body was changing, he could feel it. Breasts, he could
feel them, jiggling as he ran, hardly at all to begin with but perceptible now from
the first few strides. That's why he needed the privacy now of all times. On the
way back up that hill he could really feel them moving. Runners love that feeling of tightness that develop long into the run, as the cheeks and all the other
little bits of fat stop to wobble, but this just felt different. Unnatural, alien. The
fat on the chest was still wobbling where the hormones, the chemical castration
for his 'crime' of homosexuality were having their effects. But his mathematics
didn't leave him even then. In a letter to his friend Norman Routledge in 1952
after his arrest for gross indecency, he wrote:

> I've now got myself into the kind of trouble that I have always considered
> to be quite a possibility for me, though I have usually rated it about 10:1
> against. I shall shortly be pleading guilty to a charge of sexual offences with
> a young man. . . . No doubt I shall emerge from it all a different man, but
> quite who I've not found out.
>
> (reprinted in Hodges 2014: xxix)

Mathematics in the form of probability has reassured him that things would not
go this far, but they had misled him. His sex crime was reported in the local
and national papers. The *News of the World* headline read 'ACCUSED HAD
BIG BRAIN'. He finished the letter to Routledge with the syllogism 'Turing believes machines think/Turing lies with men/Therefore machines do not
think.' He signed off 'Yours in distress, Alan'.

It's funny the assumptions people make, that certain things are immobile,
impervious to change, and that certain things are just in your nature. But how
does your nature come about? How do they explain that? Even the simplest
fragments of nature. He often wondered about this. He had once worked on
codes, but that was many years ago, he was always interested in the mathematics underlying them, the mathematics of nature, of truth. And finding truth by
simulating it, writing mathematical codes to make a machine act like a person,

until a person, a real person that is, not a mechanical person, asking it a series of questions couldn't tell if it was a man or a machine was responding. They named the test after him – 'the Turing Test'. He wrote the mathematics behind these machines and came up with cunning tests. The 'imitation game' he called one of his papers, but that seemed like a long time ago. They say that his early work on this thinking machine helped win the war for the Allies, that it was invaluable, that it saved hundreds of thousands of lives, but that didn't stop what they did to him. They castrated 'poofters' in the fifties, with chemicals to reduce their sexual urges, and stimulating the growth of breasts, visible through his singlet, especially when he ran. The drug they used, stilboestrol, also resulted in genital degeneration.

Now he was interested in morphogenetics, the way things follow their nature, to become what nature intended, the code of nature. How does a flower develop, or a leaf? What are the instructions? What stops it becoming something else or nothing at all? What are the precise instructions, the mathematical instructions that is? He liked to run along the Bollin and see the stems of the flowers and glimpse the butterflies and the bees in all their great diversity, and the changes, as spring approached and the great push and pull of morphogenetics as things *became*. He would stop and collect some flower stems or the buds and bring them back to that house in Adlington Road at the top of the hill to study them. He was a collector now, of nature.

And yet they were saying that he could not become, that he was fixed in shape and form, eccentric, homosexual, predator, runner, oddity, genius. Sometimes he was less sure as he made his way back up the hill – about some things, at least. Others less so, others not at all. He never doubted his talent these days, just the ability of others to judge. But his belief in his own ability was not always there. He once had severe doubts.

At Sherborne School in 1927, he met his first love, Christopher Morcom, talented and to him beautiful. Next to him, Turing doubted his own abilities, comparing himself unfavourably with his friend:

> Chris's work was always better than mine because I think he was very thorough. He was certainly very clever, but he never neglected details. . . . One cannot help admiring such powers and I certainly wanted to be able to do that kind of thing myself.
>
> (from Hodges 2014: 48)

His self-doubt was compounded by the view of his teachers, endlessly criticising his 'dirty' work, and slipshod manner (Turing 2016: 44–45). The headmaster described him as 'a very grubby person at times'. The teachers' views of Turing were not flattering, even when it came to mathematics: 'Not very good. He spends a good deal of time apparently in investigations in advanced mathematics to the neglect of his elementary work. A sound groundwork is essential in any subject. His work is dirty' (School Report, Summer Term 1927: Mathematics). 'I can forgive his writing, though it is the worst I have ever seen, and I try to view tolerantly his unswerving inexactitude and slipshod, dirty work, inconsistent though such inexactitude is in a utilitarian; but I cannot forgive the stupidity of his attitude towards sane discussions on the New Testament' (School Report, Michaelmas Term, 1927: English Subjects). His school report for Latin for the same terms reads: 'He ought not to be in this form of course as far as form subjects go. He is ludicrously behind.' This endless criticism, no matter how biased with its emotional tones of disgust, 'dirty' is the word that seems to crop up in all these school reports, dirty inky fingers, unkempt, hair all over the place, must have led to self-doubt. That and the presence of Christopher Morcom, this blond-haired angelic, gifted child, as an ever-present figure; Turing paled in comparison.

Christopher, the boy he liked to collaborate with, look up to and compare himself with, won a scholarship to Trinity College Cambridge to read mathematics; Alan failed to obtain a scholarship to Trinity. He also failed his Higher Certificate maths paper (the examiner wrote: 'he appeared to lack the patience necessary for careful computation or algebraic verification and his handwriting was so bad that he lost marks frequently') although he had a certain creativity and 'knack' at mathematics (displaying 'an unusual aptitude for noticing the less obvious points to be discussed or avoided in certain questions'; Turing 2016: 52). Turing eventually won a scholarship to King's College Cambridge in 1931 to study mathematics. King's College was very much second best to Trinity for mathematics, and Alan was well aware of this.

It's all in the maths and the beauty of numbers. He knew he was different from an early age. A precocious slightly unusual mathematical talent but distinctive, messy, not immediately recognisable as such. Bullied at school, and then he had found sanctuary in Christopher, one year ahead of him at school. They shared a love of codes and puzzles and the work of the mind. They had shared a connection, a love. The one, the number one, it's not just a number. And one vacation, Christopher didn't return, he had never mentioned his illness, and now he was gone. The shock stayed with Turing forever, the location of the telling, the look of the headmaster who gave him the news, burned into his limbic system that emotional primitive system of the brain which connects the senses and memory. Images in the brain, as if written in indelible ink, dirty ink that stains. It's always assumed that the relationships of boys in English public schools are just phases, crushes, things that develop and pass in these unnaturally intense febrile environments, devoid of girls, but some things emerge as if written by a code – it's that mixture of environment and genetics that allow things to develop, to flourish – intellect, genius, love, regret, shame, pity. After Christopher's death, Turing was determined that he should *do something*, the something that Christopher had been 'called away from'. That was his driver throughout his life, it seems. Pursuing excellence on behalf of his dead friend and idol. To begin with his progress at King's was patchy. He got a second-class mark in his first-year examinations, and he wrote to Christopher's mother explaining how he could 'hardly look anyone in the face after it'. But his performance started to improve slowly at first, and then accelerating, but the self-doubt from all those teacher reports and comparison with the gifted Christopher Morcom remained. Invited to give a talk about mathematics and logic to the Cambridge Moral Sciences Club on 1st December 1933, at the start of his third year at university, he wrote in a letter home: 'I hope they don't know it already.' He was even doubting the originality of his own thinking, and his originality was all that he had.

Alan Turing was criticised by his teachers throughout his time at school, but his confidence developed year on year after that, the doubts about his own ability starting to fade. He graduated in 1934 in the maths tripos as a B-star wrangler, which meant that he had achieved first-class honours with distinction in additional papers. This was the recognition and validation that he needed. This shaped his future. The man who had the temerity later to design a machine 'brain' without even a basic understanding of human physiology – to design an intelligent brain-like apparatus governed by mathematics without the basic constraints of neuroscience. There was no doubt now in his mind about what he was doing in this world that he knew well, the world of systems and mathematical processes. In an article in *Mind*, he dismissed any counter views about machine intelligence. He wrote: 'The original question, "Can machines think?" I believe to be too meaningless to deserve discussion' (Turing 1950). He understood systems, mathematics, probability, he understood people less well be they sceptics, colleagues or those that held moderately different views. If he understood them better, and saw things from their point of view, he might have been less caustic in some of his statements.

And yet sometimes the forced removal of doubt and the confidence that grows with it can be a problem, especially when this confidence infiltrates other domains, as it sometimes can do – without any real reason or justification. Doubt can be a great inhibitor, but it can cause one to act with more reserve and caution which is sometimes necessary. So how can doubt be removed not just in a specific domain but across the board? One possible mechanism of change is a change in level of self-efficacy, which refers to an individual's belief in his or her capacity to execute behaviours necessary to produce specific performance attainments (Bandura 1977, 1986a, 1986b, 1990, 1997). Its importance to everyday life is obvious – in the words of Bandura: 'Among the mechanisms of agency, none is more central or pervasive than people's beliefs about their capabilities to exercise control over events that affect their lives (Bandura 1990: 397). Empirically, it has been relatively straightforward to demonstrate its effects. Collins (1982) showed that within each band of mathematical ability, children who regarded themselves as efficacious were quicker at rejecting the wrong strategies for solving difficult problems and they solved more problems overall. In other words, perceived self-efficacy correlates with better performance (in mathematics in this case). Feedback, even bogus feedback, produces more effort and determination in experimental tasks, but in real life personal empowerment through mastering certain tasks and experiences is the most powerful way of 'creating a strong, resilient sense of efficacy' (Bandura 1990: 402).

Turing had years of experience of mastering tasks in pure mathematics, code-breaking, mastering abstract ideas, the formulation of the computational machine and so on. These changes in self-efficacy beliefs can then impact on other aspects of how you think and feel through four major sets of processes. The first of these are cognitive processes – those high in self-efficacy believe that the environment can be influenced or controlled, they visualise success, rather than failure, which research demonstrates enhances subsequent performance (Bandura 1986a: Corbin 1972; Feltz and Landers 1983: Kazdin 1978). The second set are motivational. High self-beliefs of efficacy bias causal attributions with more emphasis on internal attributions for success ('it was all because of *me* and how clever *I* am') and external attributions for failure – they blame the situation ('the test was too hard') which maintains mood and focus. The stronger the belief in self-efficacy, the more people intensify and sustain their efforts. As Bandura points out:

> Self-doubts can set in fast after some failures or reverses. The important matter is not that difficulties arouse self-doubt, which is a natural immediate reaction, but the speed of recovery of perceived self-efficacy from difficulties. Some people quickly recover their self-assurance; others lose faith in their capabilities.
>
> (Bandura 1990: 411–412)

The third set are affective or emotional – levels of self-efficacy affect how much stress and depressed mood people experience in threatening situations. Those who believe in themselves are not perturbed by potential threats to the same extent as those who do not. They display less affective arousal in such situations (Bandura et al. 1982). The final set are selection processes – judgments of personal self-efficacy influence selection of activities and situations that may exceed their current coping capabilities, but they still 'readily undertake challenging activities and pick social environments they judge themselves capable of handling', in Bandura's words (Bandura 1990: 420).

Turing's feelings of self-efficacy grew over his years at Cambridge, Bletchley Park, Princeton and then at the Victoria University of Manchester. He may have been considered by many as an eccentric, a 'confirmed solitary', 'a sort of scientific Shelley who lived in a continual mess', but that was separate from his work and his achievements and the continual mastering of 'tasks and experiences'. And remember Bandura's words: 'they readily undertake challenging activities and pick social environments they judge themselves capable of handling.' Turing worked on the first computer, the 'Baby', built at Manchester University. It was bold and imaginative work. 'Those who believe in themselves are not perturbed by potential threats to the same extent as those who do not.' You need self-belief to be this daring, this original. Turing was working in the Coupland 1 Building at the University of Manchester. It's a bleak and soulless building – an annex to the famous Rutherford Building where between 1903 and 1919, Rutherford and his team had studied the radioactive decay of uranium-238, uranium-235 and thorium-232 in the building. The building was radioactive for decades afterwards, and those who worked there often joked about it, with a sort of gallows humour. Lecturers would bring Geiger counters to work and watch them spring into life. There was a cluster of deaths from pancreatic cancer, brain cancer and motor neuron disease, decades after Turing's death, and a trio of lecturers from the psychology department that now occupied the building assembled a report on the scientific and medical evidence. Families of the deceased blamed the deaths of contamination found in different parts of the building. Some offices and lab spaces seemed particularly dangerous. The lecturers were obviously concerned about their own and their colleagues' safety. The university had to respond to the bad publicity and commissioned an independent review of the evidence by Professor David Coggon from the University of Southampton. The report found that the cluster of deaths due to cancer was a 'coincidence' (see *Nature*, 30th September 2009; the inverted commas were in the *Nature* article). Many who worked there were not convinced. Turing would have heard his colleagues joke about the walls glowing. Professor F.C. Williams from the Department of Electrical Engineering described the laboratory in the following way:

> A fine sounding phrase, but what was the reality? It was one room in a Victorian building whose architectural features are best described as 'late lavatorial'. The walls were of brown glazed brick and the door was labelled 'Magnetism Room'.
>
> (Turing 2016: 183)

There was always a fusty smell to this building, peeling paint, torn curtains, and that brown glazed brick and that metal staircase. A minstrels' galley in the lecture theatre in the Rutherford building, hard unwelcoming seats. Nothing inviting or comfortable, always a feeling that you should not linger too long with the walls that were said to glow and the mercury from Rutherford's other experiments that had seeped into the fabric of the place. Turing had to get out even in the bleak rainy Manchester weather of December.

But then there was his personal life. His homosexual tendencies had been with him throughout. He had had several homosexual experiences back in Cambridge and elsewhere when he first spotted the nineteen-year-old Arnold Murray outside the Regal Cinema at the end of Oxford Road. It was December 1951. Murray was unemployed, hard up and on probation, as different in background, social class and position as one could possibly imagine. Turing asked him where he was going and when he replied, 'nowhere special', Alan took him for lunch across the road. He had picked up this bit of rough. After lunch, Turing told him that he had to get back to the university to work on the Electronic Brain. Arnold expressed some interest in his work, and Alan then invited this casual pickup out to his home in Wilmslow. All quite unusual for a pickup of this kind in that dodgy end of Oxford Road. Arnold accepted the invitation, but didn't show up. But Alan bumped into him again the following week and invited him once more. He came this time and Arnold stayed the night with Turing, as lovers. In the morning, whilst Alan was cooking breakfast, some money went missing from his wallet.

A few weeks later his house was burgled – a shirt, some fish knives, a pair of trousers, some shoes, a compass, an opened bottle of sherry were taken – a random and pathetic collection of oddments worth about fifty pounds in total. Alan reported the burglary to the police and two CID officers arrived to take fingerprints. Arnold's pal Harry was identified from the fingerprints. Harry was arrested and made a statement which mentioned Arnold having 'business' at Alan's home (Alan had 'loaned' Arnold some money). The police then interviewed Alan again but this time with a different focus, and he admitted that he had concealed some information from them the first time because he 'had an affair' with the gentleman involved. He then outlined three of the activities that they had engaged in. He was subsequently charged with 'Gross Indecency contrary to Section 11 of the Criminal Law Amendment Act 1885'. His punishment was a choice between chemical castration and prison. He chose the former, which meant that he grew breasts. A few years later in 1954 he committed suicide by eating a poisoned apple.

One could argue that it was the removal of doubt in the interpersonal domain that led to his ignominious fall (at least 'ignominious' in that historical and cultural context). It was replaced with this all-or-nothing intellectual daring, in which he dismissed doubt and doubters, and a self-assurance and confidence in his personal life, which may never have been warranted. His self-efficacious beliefs meant that he felt himself capable of succeeding in other strange *social* environments that he was not familiar with, where he didn't know the rules, where he couldn't read the minds of the young hustlers on the street who accommodated older men, where he couldn't see situations from their point of view. He was naïve, used and exploited. And he wasn't familiar with how the police think about men of a certain age with much younger men in their beds in early 1950s Britain in that respectable suburb of Wilmslow of all places. What was he doing inviting bits of rough from Wythenshawe out there? What might the neighbours think? Didn't he care? Couldn't he imagine their reactions?

Perhaps, he should have had some doubts about how the concept of 'an affair' might be perceived by hardened Manchester coppers of that time in that place. Self-efficacy is sometimes perceived as a panacea for our modern age with every sportsperson, businessman and woman, and every telesales agent visualising success and chanting positive words to remove doubt and build self-belief. But self-efficacy is a dispositional characteristic, which can be built and changed, and which can then generalise across domains, situations and time. It affects our thoughts, our motivations, our emotions and our choices. It removes doubt – that's the beauty of it. But sometimes we need some doubt to remain. That's what I feel when it comes to Alan Turing in the twisted, hypocritical, black-and-white world of early-fifties Britain.

Of course, some people look at the now famous Turing story as a heroic tale, a tale of a defender of truth, true to science, and true to himself, true to the right morals, right to the end. I can see that; I can appreciate that argument. But I can't read any of Alan Turing's letters from the last year of his life without feeling a sense of sadness and regret. It was an enormous loss for us all, and doubt, in this case its absence, played a significant role in it. Doubt can be a slippery phenomenon but with important consequences for the individual and society both. A great inhibitor but sometimes a *necessary* inhibitor in some contexts.

- Alan Turing began life as a somewhat cautious boy with doubts about his work and his ability.
- He compared himself unfavourably to his friend and first love, Christopher, who died very young.
- He flourished in a particular field with monumental consequences for us all – the design of the first computer.
- Turing displayed a significant change in aspects of his personality, particularly regarding doubt.
- One possible mechanism of change is a change in level of self-efficacy, which refers to an individual's belief in his or her capacity to execute behaviours necessary to produce specific performance attainments.
- In the domain of mathematics and abstract thought, his self-efficacy was high, but not in the interpersonal domain.
- The forced removal of doubt and the confidence that grows with it can be a problem, especially when this confidence infiltrates other domains, as it sometimes can do – without any real reason or justification.
- Doubt can be a great inhibitor, but it can cause one to act with more reserve and caution which is sometimes necessary.
- His self-efficacious beliefs meant that he felt himself capable of succeeding in strange *social* environments that he was not familiar with, where he didn't know the rules, where he couldn't read the minds of the young hustlers on the street who accommodated older men, where he couldn't see situations from their point of view.
- Perhaps, he should have had some doubts about how the concept of 'an affair' might be perceived by hardened Manchester coppers of that time in that place.
- Turing was naïve, used and exploited.
- He needed some doubts but unfortunately, they weren't there for him in the end.

7 Treating doubt the hard way

He was standing in front of me, me and about eleven others that is, all trying not to laugh. His hands hung heavily at his side. The gloves looked too big for him. I reckoned that he was about seven or eight, no more. Skinny white legs, a skinny white torso. He was wearing a white singlet gone slightly grey that was too big for him. It was falling off his shoulder. It was ripped in the corner. You could see the bones in his shoulder sticking out. The singlet probably belonged to an older brother. He probably went there as well. Or did. Perhaps, he'd given up and passed on the singlet and the gloves to his younger brother; perhaps he was part of our group, just standing there in that semi-circle, wondering what he could do to help his kid brother. The singlet looked old and greasy, as if it had been passed around. The room smelt of sweat, testosterone, perhaps all mixed up with traces of blood. It was a very distinctive smell that hit you as you entered, the walls were damp, there was mould in the corners. Sweat and damp, blood and damp. I've never smelt anything like that before – one of my group told me that it was the years of blood that gave it that peculiar odour. Blood in the very fabric of the place. The child was embarrassed. It was his turn. He stood in the centre of the ring. He was now in the spotlight. He knew his obligation; he knew what he had now to do. To be like his brother, to be welcomed here, to be a man.

The old grey-haired man leaning on the ropes asked him if he knew the words. The old man had a strong Irish brogue. The boy nodded sorrowfully; his head slightly bowed. He was embarrassed, more than embarrassed, he looked terrified and sad, so sad he might cry. There were a few boys his own age, perhaps friends from one of the cramped streets around here, a couple of older boys, maybe one of them really was his brother, and then some toned professional boxers: two were familiar faces, one was a world champion. The media called him 'Motormouth', the most arrogant world champion in the whole sport of boxing. Worse than Muhammed Ali ever was. He was slightly built, like the boy himself, but Arab looking. He had been going to that gym since he was seven years old. He couldn't smell that gym anymore: he was used to it. The old grey-haired man was his surrogate father, that's what everybody said. The white skinny boy glanced at the world champion as if for approval. The world champion just laughed. 'Get on with it,' he said, glancing around for a laugh.

DOI: 10.4324/9781003282051-7

The boy opened his mouth, but no sound came. 'I've forgotten the words,' he said quietly. 'I can't do it. Why do I have to do this?'

'Everybody does, there's no exceptions, even your man over there,' said the old grey-haired trainer. He gestured with a shrug of his left shoulder to the world champion.

'Too right,' said the champion. He rolled his shoulders and glanced around again.

Cocky is how I would describe it. He was still laughing, as was his friend standing next to him, adding to the obvious embarrassment of the boy. This was what this boy wanted to be – confident, cock-sure like them, rolling in dough.

The old man started it off for him then the boy started singing, quite quietly, almost whispering, it sounded quite tuneful.

'Red and yellow and pink and green.'

There was a pause. Nothing happened. Everyone was staring at the boy in the centre of the ring. The young boy tried, he repeated the words 'red and yellow and pink and green'. There was no tune, he was just speaking the words back to the trainer.

'Sing,' said the grey-haired man.

'You have to sing,' said the champion. 'Anybody can just speak the words. We're different down here. We're not anybody.'

The grey-haired man continued singing: 'Purple and orange and blue.' The boy started to join in, he looked as if he was going to cry, but they were now singing together. 'I can sing a rainbow/Sing a rainbow/Sing a rainbow too.'

I'll sing the next verse said the trainer. He was still in tune. 'Listen with your eyes/Listen with your ears/And sing everything you see/I can sing a rainbow/Sing a rainbow/Sing along with me.'

'Now it's your turn with that first verse again,' said the trainer. This time the boy sang on his own. He had quite a sweet voice, very childlike. Everybody looked surprised. It sounded innocent. There was loud applause when he stopped and some whooping.

The skinny boy in the centre smiled widely, I noticed that his right leg was trembling a little.

'Can we have *my* music on again now,' said the champion, 'and get back to business.' There was stress on the first syllable of 'business'. Loud rap music boomed out: it made me jump, it had a visceral feel, it sounded like a nightclub kicking into gear. That music and the smell combined to give it this primitive, almost animalistic feeling, which contrasted sharply with the nursery rhyme just sung in this child-like way, as if we'd just stepped into a different room, or a different world where all innocence had gone.

'I call this music "crap rap",' said the trainer to me. He glanced at the champion, Naseem Hamed or 'the Prince', as he called himself. Naz ignored him. The boxers started walking in small anti-clockwise circles in the ring. The Prince, his heavily muscled black friend, nearly a foot taller than the five foot four–inch champion, and the rest of this motley crew, including the new boys, all walked in this slightly solemn line. The champion swayed to the music, the rest didn't. He seemed to think that he was dancing in front of a mirror. I noticed that an older boy was patting the young boy on the back. He looked like the younger boy, similar colour hair, similar build, almost malnourished look-ing, definitely poor. The new boy smiled and followed the other boxers around the ring. He didn't look quite so terrified now. He had passed the first test.

This was Brendan Ingle's gym in Sheffield a few decades ago. Brendan was one of the most successful boxing trainers of all time in the UK. This gym of his in Wincobank with that acrid smell and the mould on the walls and ceiling had produced over thirty major titleholders, five world champions, six European champions, six Commonwealth champions and fifteen British champions since first opening its doors in the 1960s. The nursery rhyme was part of his approach: it was all about teaching confidence, all about eradicating doubt.

Doubt and self-doubt are a central aspect of all sports centring, of course, on fear of failure (Smith at al. 1990). But this has several strands – fear of failure itself and the shame and embarrassment associated with this failure (McGregor and Elliot 2005), but also fear of the social consequences of poor performance and worry about not living up to the expectations of one's trainer, family and friends (Correia and Rosado 2018). You can lose your self-worth in that ring, you can lose your livelihood, your friends, your routines, your support network, your self-respect. You can even lose your life. Brendan liked to remind them of this. Those athletes across sport with the greatest fear of failure display greater anxi-ety levels generally – such athletes worry more frequently about factors relating to poor performance and the resulting social and psychological consequences (Dunn and Dunn 2001). All possible failures are encoded into a set of negative emotional signals, which we would describe as fear, shame and embarrassment. These signals give the experience an immediacy. McGregor and Elliot (2005) demonstrated this conceptual link between fear of failure and shame. 'For indi-viduals high in fear of failure, achievement events are not simply opportunities to learn, improve on one's competence, or compete against others. Instead, they are threatening, judgment-oriented experiences that put one's entire self on the line' (2005: 229). These emotions often trigger avoidance behaviours – those with high fear of failure often adopt specific avoidance-based goals and 'self-handicapping strategies' that have negative effects on performance and psy-chological well-being. This is the challenge faced by that old grey-haired man leaning over the ropes. He must stop that skinny kid losing the fight before he even steps into the ring – 'self-handicapping' is how psychologists describe it.

There is evidence that fear of failure is passed down through the generations. Singh (1993) found that negative maternal characteristics such as irritability and dependency are associated with high fear of failure in children. Greenfield and Teevan (1986) reported that the father's absence from the home, especially due to the father's death, was also related to higher levels of fear of failure. Elliot and Thrash (2004) showed that both mothers' and fathers' fear of failure was a posi-tive predictor of performance–avoidance goals in their children. In other words, it is passed down through the generations. They point out that although fear of failure is an achievement motive, its conceptual grounding in the experience of shame means that it is inherently relational, as it 'involves an awareness that this defective self is exposed before a real or imagined audience, is judged unworthy of love, and is in danger of being abandoned' (2004: 958). Fear of failure affects how the parents view failure and their own failures, and therefore how they respond to the failures of their children. The parents self-handicap in their own lives, and this is passed on to their children. Better not to try than try and fail and suffer all that shame and embarrassment. This is what goes on in some people's minds and is the mechanism of transmission. This trainer in this most humble part of Sheffield must break this chain of transmission. Families pass on their fear of failure to their children – Wincobank was full of families who no longer tried; Brendan Ingle had to block this process.

Brendan took the young Naseem Hamed under his wing when he was just seven years old. He saw this kid fighting in the playground of the local primary school and thought that there was something about him – he held his own, he was fighting three White kids all at the same time, but he didn't back down. Brendan thought that he was a little Pakistani kid (Naz's parents are from Yemen) and the fight might well be race related. The National Front were big in the area at that time. Brendan could see that this kid had guts. Naz lived just up the street and started coming to the gym every day. Brendan took him everywhere with him – this slight Yemeni boy with wavy dark hair trailing behind this old trainer. He wanted the kid to see the world of boxing first-hand – the good side and the bad. Everyone talked about their father-son relationship – they were inseparable. A surrogate father to teach confidence and life skills, to instil achievement motivation, to abolish the fear, to get this boy to aim for the sky – 'then we will both be millionaires', Brendan would always say. And they would both laugh.

Of course, boxing is not just any sport – fear is always there. You stand in that ring before the fight, disrobed, stripped almost naked, all the trapping of show business (the ring walk, the music, the image) now behind you, looking directly into the eyes of another man who wants to hurt you badly. Joyce Carol Oates wrote that boxing, unlike wrestling, is 'real', not theatrical, rehearsed or simulated (Oates 1987: 106). Your opponent is trying to outstare you; you don't dare break eye contact. The arm of the referee separates you, but this one safety barrier will soon be removed. Brendan liked to remind all his fighters that boxing is the only sport where you can be legally killed. There was a sign up in the gym with the same message. You stare into your opponent's eyes, trying to stop any flicker of emotion. This takes effort. If you have any doubt, then your opponent will read it. He can smell your fear. He watches your eyes, and one single micro-expression can tell him all he needs to know. As Oates wrote, 'No two men can occupy the same space at the same time. . . . Boxing is for men and is about men and is men' (Oates 1987: 72). You can smell your opponent's breath; you're that close. You know what's going to happen once the referee's arm is pulled away. His breath smells nauseous. It's only a matter of time now.

And then that doubt pops into your head – how much could it hurt? Many of the boxers that arrive at Brendan's gym have been bullied when they were younger. That's why they'd gone in the first place – to learn to take care of themselves (Woodward 2004); they know all about fear. They've had their doubts before. You're going to try your best, but you'd only been called for that fight that afternoon. Nobody expects you to win. You would never dive, but how many rounds are the punters expecting? They've paid good money. Even the support fights must be good. Your opponent thuds his gloves together. Droplets of sweat fly off and land on you, caught in time. How many rounds of this? The bell sounds. You almost choke on your saliva. You begin. He lands the first punch to your ribs. You weren't expecting that – a softener, a punch that will soften you up – physically and psychologically.

To train a boxer you must deal with fear, you must teach them to rid themselves of doubt and control their talk, their behaviour, their emotions – you have to teach them to be confident and self-assured. That's why so many successful boxers appear arrogant. They've learned to conquer fear, unlike the rest of us, sometimes paralysed by fear. You must train them over many years in self-efficacy until they feel that they can do anything they want, beat any opponent, conquer the world. These trainers are the experts at getting rid of doubt. Brendan's approach was always very distinctive and it worked – based around performance practices – hit and don't get hit (Herol Graham, an earlier Ingle boxer, who was one stoppage away from a world title would stand in the ring with one hand behind his back and invite punters in a local working men's club on a Sunday afternoon to punch him on the head – for a wager – they never could), bag work, pads, sparring, of course (but with no shots to the head) but with endless other routines – footwork and 'doing the lines', marked out on the floor of the gym, acrobatic skills, and somersaults into the ring plus these odd narration skills and public recitals of nursery rhymes.

Why did he insist that they recite nursery rhymes in public? And not once or twice, or just when they arrived, but many times over the years. He made them do this to overcome their fears about public speaking, to overcome their anxieties about making a fool of themselves, to teach them to be less self-critical, to control their thoughts and their emotions. I had gone to that gym back in the 1990s to observe this process of training in action and to see its effects – it was ethnographic research (Beattie 1996, 2002). I took up boxing to fit in with the group. I relished the idea. My Uncle Terence had been a good amateur boxer in his younger days in Ligoniel in Belfast, and as a child he would let me box him on a Saturday night when he and my father got back from the pub. Spot, our black and white fox terrier would only ever attempt to bite him on a Saturday night, when he reeked of Guinness. Spot would be on his shoulder biting his neck, I would be clambering over the back of the settee boxing his ears. One night I gave my Uncle Terry a black eye. He was very proud of me for that. He would routinely drink fourteen pints of Guinness on a Saturday night, so I don't think he felt it. He didn't even flinch, although he threw Spot across the room, and Spot was the innocent party. We always worked ourselves up into a great lather on a Saturday night, all salty sweat and sickly foam from the dog. The dog would have to go and lie down in front of the electric fire afterwards to cool down, and then it would sneak round the back of the settee to drink the porter out of my uncle's glass. That was all a lifetime ago, but I thought that it would stand me in good stead for the ring; I thought that I could punch. I trained with a group of doormen; a couple of them were ex-boxers, including Mick 'The Bomb' Mills. I liked Mick, always funny and welcoming, and hard as nails. I enjoyed the bag work and the running, but Brendan always looked alarmed when he saw me step into the ring with them. 'Don't let them use you as a punch bag, Geoffrey,' he warned.

But the sparring and the gym work allowed me to get close enough to observe the training and the change in the character of the boxers in that gym in more detail. That was what I wanted to see – psychological change and how this is embedded in relationships, and how doubt and its eradication depends on social processes, sometimes long and protracted. I made many observations:

> They were standing in his parlour in front of a pair of scales. This grey-haired man in his fifties stripped to his underpants and a small, slight Arab boy. On one side of the room was a large bookcase. There were a lot of books on Irish history. Black covers with the green of the shamrock. Serious reading. It could be the parlour of a minor academic or a priest. On the other wall was a large, framed photograph of the boxer Herol Graham, the Graham of years ago, the Graham of eternal optimism and promise, the most famous boxer from Sheffield, the man who nearly became the champion of the world. Nearly. The grey-haired man got on the scales first. 'Twelve stone dead. Now it's your turn, and don't forget I've been warning ye.' His Irish brogue was as thick as buttermilk. The small Arab boy stepped forward in a mock swagger. 'It'll be all right, Brendan, don't worry.' His was a cocky Sheffield accent. 'I'm young, fit. I *am* the business.'

Brendan Ingle averted his gaze from the scales just for a moment. He felt that he needed to explain:

> The problem with Naz or Prince Naseem Hamed, as he's known in the ring, is that he knows he's good. At that age it's bound to go to your head. He loves himself and why not? I've had the lad since he was seven; he's nineteen now. His father is from the Yemen. I was passing on this bus up the road here, and the bus stops outside a school. It's three o'clock and the school is just getting out. There's this little kid who I thought was a Pakistani pinned up against these railings fighting these three white kids off. All three of them are kicking and punching at him. My first thought was that life doesn't change. I can remember getting into scraps in Dublin when I was a kid. Sure, they always had some reason to pick on you for a fight – 'You're not proper Irish – your grandfather's English.' 'What kind of a name is Ingle, for God's sake?' They could always find something. And if it wasn't that it would be 'You think you're great because your brothers are boxers.' I was from a big family – I had ten brothers and four sisters. But I was impressed with

this young Pakistani kid. I can spot talent a mile off. I'll go anywhere where there's a fight. I'll get me distance and I'll watch. I won't get involved but I'll watch them. I've always been interested in what goes off in confrontations between people whether they're arguments or fisticuffs. I'm interested in human nature, and that's what you see when people fight and argue. I like to see what starts trouble and how it goes on. I like to watch who's involved, who makes themselves busy, who's the matchmaker, and who's stirring the whole thing up. I watch and I learn. I could see this young Pakistani kid had talent. I ran home and told my wife Alma. I was right, as well. Naz has won seven British titles, and he's boxed for England as an amateur. Now he's a professional but even as a newcomer he's earning good money. But it's still only the start, now it's all about self-discipline.

Naz climbed onto the scales.

These scales never lie, Naz, remember that. . . . Eight stone nine. What did I tell ye? What did I say? You were boasting about all the crap you'd eaten yesterday. Fish fingers and chips. Well, this is what you get. You're three pounds over the weight. I've said to you time and time again – you're eating all wrong and you're sleeping wrong. You're up far too late playing snooker.

'But, Brendan, I'm beating everybody. I'm knocking them all out. You know how good I am.' Brendan replied:

You may be the greatest thing since sliced bread, but this is your comeuppance. Three pounds overweight two days before a big fight can be hard for any fighter to shift let alone a bantamweight. You've got less than two days. This will be a test of what you're made of outside the ring.

'I always knew that young Naz would fail to make the weight someday,' said Brendan, exactly one month later:

That lad could eat for England. I didn't allow him to eat anything for the remainder of the Monday or on the Tuesday morning. At lunchtime on the Tuesday, I weighed myself again – I was twelve stone again. I explained to Naz that by the time we got down to London, with the stress and strain of me driving all the way, I'd have lost two pounds. It wouldn't be so easy for him. We were staying in a flat above the Thomas à Becket gym. I'd brought my scales with me. When we got to London my weight was down to 11 stone 12 pounds, exactly as I'd predicted. Naz had lost a pound, but he still had two extra pounds to shift. Naz's room was cold. I took the blow heater out of my room to give to him. I also switched on the sunbed in the corner of his room. It's my job to see he's as comfortable as possible before a big fight. I explained to him that he'd lose nearly one pound sleeping. We were both starving. I hadn't had anything to eat or drink all that day either. It's no good me trying to motivate or inspire somebody else if they're trying to make the weight and I'm eating.

It was a rough night. The bed was damp, and somebody down below was playing some old Beatles records. By the next morning I was down to nearly eleven stone eleven pounds, I'd lost almost another pound. Now I was counting in ounces. I went to the toilet, and I was now only six ounces over the weight I'd set myself – I had to lose three pounds just like Naz. Naz was less than a pound over the weight. So, I went to the toilet again. Naz accused me of having something to drink. He said that I must have sneaked a drink. I kept going to the toilet and my weight kept dropping. Naz couldn't believe that I was going all this time without having something to drink. But the proof was in the scales. He could see the ounces coming off. He knew I wasn't cheating, and that he and I were going through this thing together. But every time I went to the toilet, I had to reassure him by stepping on the scales again to show that the weight was coming off, and that I wasn't sneaking a drink. We were watching each other like frigging hawks. I took him for a walk and by the time we got to the weigh-in he was half a pound under the weight. He hadn't had anything to eat or drink since the Monday, that's thirty-six hours without anything. I took him to a restaurant after the weigh-in, but he couldn't finish his soup or his spaghetti. His stomach had shrunk. But he felt good. I told him that day that he'd grown in my estimation. Naz has this little routine when he gets into the ring. He jumps over the ropes, just like Chris Eubank, but then he does a flip holding on to the ropes and then three flips across the ring. He's a bit of a showman. But I told him that night that I just wanted one flip from him, then he was going straight to work. I told him that he was going to mentally and physically destroy his opponent. Incidentally, his opponent was unbeaten before that night. Naz had him down three times in the first round. He knocked him out in the second. I told him afterwards that there is nothing to stop him becoming world champion.

'Who's going to stop you now?' says I.

'Nobody,' says he.

'Right,' says I.

Brendan Ingle always had a certain way with the boys. He must have had. They came from all over Yorkshire to train and spar in his gym in Wincobank, although most, it must be said, came from the streets roundabout. At that point, this small gym had by then produced four British or European champions, and almost a world champion in Herol Graham. Then Johnny Nelson had gone the whole way (or part of the way, depending upon who you listened to, because most fight fans seemed to regard his recent acquisition as a bit of a Mickey Mouse title) and taken the WBF world cruiserweight title from Dave Russell in Melbourne. The title was not recognised by the British Boxing Board of Control. It was the Irish blarney they were all saying. 'The blarney did all that?' I asked incredulously.

Matthew was twelve, with a round, open face. Brendan was perched on some wooden steps by the side of the ring. He called Matthew over. 'How long have you been coming to the gym?' 'Three years,' said Matthew. 'Tell Geoffrey what it was like before you came to the gym.' 'It was terrible, I had no friends, and I was being bullied all the time at school.' 'Tell him what it's like now,' said Brendan. 'It's great. I've got lots of friends, and I'm not bullied now.' Brendan squeezed his arm tighter and said:

> This lad here can't fight. He'll never be able to fight. I'll teach him to dodge a bit in the ring. I'll teach him to mess his opponent around, to make the other fell look bad. I'll build up his confidence. I taught him that if anybody comes up to him on the street and starts to bully him, that he should just should 'Piss off,' and run away. I'm teaching him personal and social skills for life. My job is to get these lads through life as safely as possible, both inside and outside the ring.
>
> It can be rough around here. I was the one to bring the first blacks into this area to my gym. So, the National Front put posters all over my house and scrawled their name on the walls of the garages outside the gym. You can still see their graffiti to this day. But they've gone, and I've survived, and that's what it's all about. I knew this guy who was big in the National Front, and he ended up marrying a black girl. So, it was all a load of bloody bollocks anyway.

In the ring above us there were five boxers. Two were Black. One was a powerfully built novice boxer, the other was Johnny Nelson. Two were small White boys, probably no more than ten, the fifth was a serious-looking Asian youth, dressed in a black polo and black tracksuit bottoms, who stalked his opponent before unleashing incredibly ineffectual-looking punches. They took turns at the sparring with each other, with one always left out. Johnny Nelson with the muscled Black novice, then Johnny with one of the boys. 'Only body shots up there. I won't stand for any boxer in this type of sparring giving his opponent one accidentally on purpose like to the head,' said Brendan. '*Time!*' he shouted, and all five boxers walked slowly around the ring in an anti-clockwise direction. 'In my gym the professionals train with the novices. They can all learn something from each other. Bomber Graham used to stand in the middle of this ring and the lads would try to land a punch on him. They never could. *Change over!*' The boxers touched gloves gently, as if they were in some barn dance, and started again with a different partner.

A sixteen-year-old stood by the ring bandaging his hands. Brendan called him over. 'How old were you, Ryan, when you came to the gym for the first time?' 'Six.' 'What did I say to you?' 'Do you know any swear words?'

> So, I got him to tell me every swear word he knew: 'fuck', 'bastard', 'wanker', the lot. So, says I to him, 'From now on you don't swear when you're in this gym and you do as you're told.' It took him by surprise, you see. Then says I to him, 'What do they say about the Irish where you come

from?' and he says, 'They're all thick bastards.' But this thick bastard says that this lad will be winning a gold at the Olympics in three years. When the English shout 'Fuck off, you thick Irish Mick' at me, I just remind them that they were riding around on dirt tracks before the Irish came over.

He pulled out a book that he has been carrying around in the pocket of his anorak to show me. It has pages of nineteenth-century political cartoons on the Irish problem.

This shows you what the English thought of the Irish. The Irishman was always portrayed as a wee monkey. Here's the Irish Guy Fawkes, a wee ugly monkey in a hat, sitting on top of a keg of gunpowder before setting light to it. The thick wee monkey bastard. But the Irish are too cunning for the English. I've had the British super-middleweight champion. When I was starting him off I called him Slugger O'Toole. The Irish are great boxing fans, so I reckoned that they'd turn out in force to see an Irish boxer with the name O'Toole. Slugger would come into the arena dressed in green, and it wasn't until he'd taken his dressing gown off that they would see that he was black. So, they'd all be shouting, 'He's not Irish!' and I'd say 'What's the matter with you? Haven't you ever seen a Black Irishman before?' And then when they asked me his real name, I'd reply, quite truthfully as it turns out, 'Fidel Castro Smith'. They'd not believe me anyway.

Brendan pulled Ryan close to him.

'Who's the only person who's ever going to lick you?'
  'Me, myself.'
'Who's responsible for you?'
  'Me, myself.'
'Correct.'

This was the routine they had rehearsed many times. Ryan knew when to come in, and he knew all the unvarying responses. It was like the litany from a church service.

Some people think it's easy to be a boxing trainer and manager. You just cream off your twenty-five per cent and the lads do all the work. But that's not how it is. I take these lads in when they're kids, and I have to work on them. I have to build up their confidence. I have to teach them about life and replace all the crap they've learned. They come here and their heads are full of it.

He called over a somewhat shy-looking boy with a thin moustache and dull greyish shorts. His girlfriend had been sitting in the corner of the gym biting her nails all afternoon.

'What school did you come from, Matt?'
'Arbourthorne.'
'What kind of school is it?'
'Special needs.'
'What were you there for?'
'Because I was a thick bastard.'
'What are you now?'
'A clever bastard.'

The responses were instantaneous, starting almost before the question fin-ished. He pulled Matt closer until their faces were almost touching.

'Who didn't you like when you first came here?'
'Pakis.'
'Pakis and Blacks?' asked Brendan.
'No, just Pakis. I always thought that the Blacks were all right.'
'Who don't you like now?'
'Nobody.'

'When that lad came here, he had nothing going for him,' said Brendan.

Now he's part of a team in here. There are a lot of Pakistanis in the team. He trains with champions. He'll be sparring with Johnny Nelson in a couple of minutes. I can identify with these lads. When I was a boy I had what you would call now 'learning difficulties'. Over in Ireland I was just a 'thick bastard'. I struggled with my spelling, my reading and everything else. I struggled with Latin and Gaelic. I can still say in Gaelic 'What is your name?' 'My name is Brendan Ingle.' 'Where do you live?' 'Dublin City.' 'Do you have any money?' 'No, I haven't.' I can recite all these verses in Latin, but I've no idea what any of them mean. These verses were beaten into me. One of the nuns was a right bastard with her leather strap. I was beaten because I was thick. But this taught me a valuable lesson: you can't change people's attitude by mentally or physically abusing them. The only way that you can change people is by engaging in dialogue with them. Only dialogue.

He shouted over to Matt. 'Who is the only person who can beat you?' 'Myself,' replied Matt across the crowded gym. 'My lads come to me with all their problems. I always say to them that if you haven't killed anybody, then you haven't got a real problem. I can sort everything else out.'

He turned to the boxers in the ring. 'Now lads, before you get down out of the ring, I want to see you one by one jump over the ropes.' Johnny Nelson did it with some flair, the rest struggled to get over. 'I let them jump off the second rope, if they can't manage it,' said Brendan. 'I do it to build their confidence in all aspects of life. Boxing isn't just about punching and how to slip a punch. It's about building confidence and learning to survive.' He shouted up to Matt, now sparring with Johnny Nelson. 'Matt, what do you say if some dirty pervert comes up to you in the street?' 'I shout "Fuck off!"' 'What do you do then?' 'I run like fuck.'

'I'm teaching them how to survive inside and outside the ring,' said Brendan, 'which isn't that bad for a tick Mick.'

Brendan Ingle used to call himself a professor of kidology, it wasn't just for my benefit, but he really was a maker of champions. Naz won the world title from Steve Robinson on the 30th September 1995 at Cardiff Arms Park. According to John Ingle, Brendan's son and Naz's trainer, Naz won the fight at the weigh-in.

> It was like Muhammad Ali predicting that he was going to knock the bear, Sonny Liston, out in five rounds. Everybody used to laugh except the guy he was talking about. Naz got to Steve Robinson at the weigh-in. He was looking straight into his face and saying, 'Steve, you're going to get beaten.' It wasn't done in a nasty sort of way; it was just done in a matter-of-fact way. The cold stare of Naz can be quite intimidating. I could feel Steve Robinson crumbling inside as Naz was saying it. I said to Johnny Nelson, 'Look at that, he's gone already.'

Naz was then twenty-one years old and the champion of the world. Brendan Ingle had been training him in his run-down gym in St Thomas' Boys Club for fourteen years. Of course, Naz had exceptional talent, but there was something else – there was that special bond between them, like father and son, and this intensive boxing training, and psychological training – the removal of doubt, over all of those years by the professor of kidology, someone who used his intuitive skills, shaped by years of experience, to mould the body and the mind of a champion. I watched Brendan planning it all, the road to the world championship, getting Naz to practice his entrance to the ring and that iconic and arrogant somersault over the ropes. The flip over the ropes was part of the routine that Brendan made all of his boxers practise in the ring (even those who would never have a professional fight). It was part of his confidence-building exercises. And it worked; I saw it happening before my very eyes over the months and years I spent in that dusty gym. All the boys knew that the fighter's entrance to the ring was a critical psychological moment.

I always recall David Remnick's description of how Floyd Patterson entered the ring in 1962 to defend his world heavyweight title against the challenger Sonny Liston in his classic book *King of the World*.

> He bent through the ropes and into the ring, but he did it stealthily, nervously, with quick glances all around, like a thief climbing in a window on the night he knows he will be arrested at last. He was in a terrible state. His eyes flicked around the ring. Rarely had fear been so visible on a fighter's face.

Liston won, of course; Floyd Patterson was knocked out in the first round.

Brendan always knew that the way a boxer entered the ring could leak a lot of information about his internal emotional state. The flip over the ropes was the greatest mask of them all. The fighter could be churning up inside and his opponent would never see it. As long as you got it right, and Brendan saw to that. It required concentration and hours of practice to carry it off. The practising of the entrance to the ring was as routine down in St Thomas' as sparring itself. Importantly, perhaps critically, the flip over the ropes distracted the fighter himself from what was going on inside. It allowed him to deal with the doubt inside. He was concentrating on something other than his inner worries.

I remember one afternoon in the gym watching Naz being taught a new, more eye-catching somersault. 'Get your arms straighter,' said Brendan, 'get a bit more spring into it.' Brendan had a great idea at the time that the Arab Prince should make an entrance on a flying carpet, courtesy of Paul Eyre's carpets, who were big in South Yorkshire. Ali Baba comes to boxing. It was all being mapped out on a little scrap of paper on the dusty steps at the side of the ring. In the drawing, Naz was sitting cross-legged on the Paul Eyre's carpet. Naz pointed at the drawing. 'Where's the fans, Brendan?' he asked.

'Oh, that's them down there,' said Brendan, shading the bottom of the page with the thick lead of the pencil. It was just a great black smudge. 'There's hundreds of them just waiting for "The Prince" to arrive.'

'Draw us a few more, Brendan,' said Naz.

And Brendan shaded a bit more of the paper. 'Is that enough for you? Or do you want a few more, you greedy beggar?' asked Brendan, and they both laughed.

'How do I get down off the carpet?' enquired Naz nervously.

'Oh, we'll worry about that a bit closer to the time,' said Brendan. He folded up the bit of paper and put it in his pocket, and left Naz and me to talk some more.

Naz always recognised its importance. He said to me, 'Well, if somebody did it to me, I'd think, "Well, that is a confident man. I just hope he can back it up."'

But it was the verbal routines that really stuck out that he used on all the lads. We heard him with Matt earlier. Over and over, it went. Matt's responses built into a conversational routine, until it was his automatic response. 'What are you now?' . . . 'What are you now?' . . . The other boys would be told to gather around to listen to Matt's responses, Naz included. Then it would be the turn of another member of the group, then Naz's turn. These were the sounds of that gym with the high ceilings and the sharp acrid smells of sweat and testosterone, gloves slapping on hard bodies, the squeak of the wooden floorboards with fast, shuffling feet, boys standing upright, reciting liturgies about themselves, liturgies about their beliefs and their aspirations, as the others stood in a reverential silence. 'I am a clever bastard.' 'I won't let anybody bully me.' 'I will be the champion of the world.'

Brendan Ingle, with no formal education, from the slums of Dublin, soaked in Roman Catholicism, understood the power of the word, and of the group, and the nature of group influence. He also understood the power of getting these lads to articulate these things – things that they did not actually believe about themselves ('clever', 'non-prejudiced', 'clever enough to survive on the street'), or their upcoming opponents ('nothing compared to me', 'a muppet', 'a nobody'), or their destinies ('world champion in the making', 'the new Naz', 'a millionaire with all the trimmings'), but they needed to say them, over and over again, so that one day they might actually believe them. Did Matt really believe that he was 'a clever bastard' at that time? Of course not. But, I watched him say it so many times, in front of his friends and training partners, that I sometimes wondered whether he may have started believing that he was, at least, 'a clever bastard', and certainly no longer 'the thick bastard' that had turned up lost and alone at that gym. But that's not quite the same thing, far from it.

Where did Brendan pick this up from? Perhaps, partly from his Roman Catholic upbringing. 'Forgive me, Father, for I have sinned', at the start of every confession, saying that you had sinned because that is what is required, even when you couldn't think of one. But if you say it enough times of your own free will (and who's actually forcing you) then you must have sinned. You start believing it, so you begin that search for your less apparent sins, sins of the heart, sins of omission, sins of impure thought. Brendan's Catholic background certainly found its way into the tone and automaticity of these verbal routines. But, of course, it was also partly from Cassius Clay, 'the Louisville Lip', ever since he had gone bear hunting in 1964 with his bus with the slogan 'CASSIUS CLAY ENTERPRISES. WORLD'S MOST COLORFUL FIGHTER' and 'SONNY LISTON WILL GO IN EIGHT'. He called his opponent Sonny Liston, the man who had knocked out Floyd Patterson in the first round, the 'big ugly bear'. He wrote poems about what would happen in the fight, as read on CBS' *I've Got a Secret* (1964):

> Clay comes out to meet Liston
> And Liston starts to retreat
> If Liston goes back any further
> He'll end up in a ringside seat.
> Clay swings with a left,
> Clay swings with a right,
> Look at young Cassius carry the fight.

He predicted how long the fight would last. As Norman Mailer commented before the fight (see Kempton 1964), if Clay were to win the heavyweight crown, then 'every loudmouth on a street corner could swagger and be believed.' Being a loudmouth was no longer a good indication that you couldn't fight as well. Clay understood the power of language, in this case playing the fool in front of Sonny Liston. He said 'Liston thinks I'm a nut. He is scared of no man, but he is scared of a nut.' Ergo, Clay is saying, he is scared of me. Clay later admitted that he had only ever been truly afraid only once in his career, and that was in this first fight with Liston. 'That's the only time I was ever scared in the ring. Sonny Liston. First time. First round. Said he was gonna kill me.' Cassius Clay had found a way of dealing with his doubt and his fear. Floyd Patterson, who talked freely about his fear, indeed more freely perhaps than any other boxer in the history of the sport, perhaps did not. Patterson articulated that fear:

> A prize-fighter who gets knocked out or is badly outclassed suffers in a way that he will never forget. He is beaten under the bright lights in front of thousands of witnesses who curse him and spit at him. . . . The losing fighter loses more than just his pride and the fight; he loses part of his future. He is one step closer to the slum he came from.
>
> (Patterson 1962; cited in Patterson and Gross 1962)

When Cassius Clay became Muhammad Ali, he told the world and himself enough times that 'I am the greatest!' so that both we and he came to believe it. His shining charisma derived in large part from his self-belief. He seemed invincible, and for years he was. This was the power of language for changing history by managing doubt and determining the belief structure of the individual to make that possible. This was not lost on the twenty-four-year-old Ingle. His own boxing record was mixed with seventeen wins and four losses. He didn't try these techniques on himself, but he could train others to use them. The seven-year-old Naseem Hamed in the corner of the gym, who was telling anyone that would listen that he would one day be the world champion, until everyone was bored with it. Later, in the build-up to the world championship fight, Brendan would say, 'The greatest thing since sliced bread. They used to call Muhammad Ali "The Greatest". Wait until they see what Naz is going to be doing over the next few months.' The tabloids called him 'Megamouth' or 'Motormouth'. They were saying that he had a level of ambition and plans for world domination only seen before in a Bond villain. Some called Naz confident; most called him arrogant. He belittled his opponents with his showboating. He told them to their face that they didn't have a chance.

A few years later, I interviewed Naz after he had become world champion in a plush hotel in London, a setting a million miles in money and social class from the musty Wincobank gym. He had changed but seemed pleased to see me. I began by asking him if he had ever been frightened before a fight. His look said it all.

> No, I've never been frightened before a fight. You should see me in the changing rooms beforehand. I'm totally the opposite. I get all the guys in, all my entourage from the gym. I pack out the changing and slam some music on, hip-hop, rap, whatever my mood desires. We'll be having a great laugh in there, and I mean a great laugh. I'm talking about laughing, giggling, cracking jokes. This is five minutes before the fight. As long as I've got bandaged up and oiled up, I'm happy and ready to go. As soon as they say, 'The television is ready, you're on,' I'm a different person. I'm ready

to walk out and I'm ready to do the business. There is nothing else in my mind except to go out, get into that ring in style, as you've seen me do, and take an opponent apart and take him apart in style. In style!

(Beattie 1998b: 54)

I then asked him when was the last time that he was frightened of anything. We must remember that this was a boy who had been going to that gym since the age of seven, overcoming fear of public speaking with the nursery rhymes in front of his fellow boxers (and any journalists, politicians or celebrities who happened to be passing), practicing his somersaults into the ring, practicing control of his thoughts, emotions and behaviours. His response could not have been clearer. 'I can say that fear really never gets to me' (Beattie 1998b: 54).

And what about 'doubt'? Did he have any doubts in his mind that he might lose his world title fight against Steve Robinson in Cardiff? 'No, there was never any doubt in my mind. From day one, I knew that I was going to be in there with my perfect style. I knew that he was there to be taken' (Beattie 1998b: 57).

So here was a man schooled in a particular way that removed both doubt and fear. My years in that gym taught me something about the intensity of that training. Of course, like everyone else I always wondered what might happen if Naz were to lose a fight. Many boxing fans looked forward to such an occasion with great relish. And it came to pass in the MGM Grand Garden Arena in Las Vegas, Nevada, in April 2001 – his opponent was Marco Antonio Barrera. Naz had held the WBO world featherweight title for nearly five years; he was guaranteed six million dollars for the fight. Barrera, the underdog, won a unanimous points victory. I then travelled back to interview Naz in his own private 'boxing palace' on Abbeydale Road in Sheffield. Brendan and Naz had split up years previously, before the defeat to Barrera. There were framed photos of Naz all over the walls, and framed pairs of shorts, including the leopard skin shorts with 'Prince' on the front that he had worn the night he won the world title. I pointed out to him the date beneath the framed shorts was wrong. It read 30 September 1985, not 1995. 'We never noticed that,' said Naz. There was a cushion cover with Naz's portrait on the front hanging on another wall. I stood there looking at the ring in the centre of the marbled room and then at this huge, polished gallery of images of the Prince from every possible angle. His brother Riath nudged me. 'You can see that his ego has landed here with all these pictures.' Naz jumped in. 'What else could I put up? I wanted to make the gym a little bit more homely.' It was an odd word to describe a gym with endless pictures of yourself.

So, what happens when you come back down to earth with a bump, even with all this psychological coaching? I asked directly. 'Tell me about losing to Barrera,' I said, 'How did that feel?' I noticed that he moved away from me slightly.

> I knew that I had lost before the last bell went. . . . I don't want to make any excuses but there were loads of things going wrong. Everything that could

have gone wrong went wrong that night. My bandages, my gloves. I mean
I didn't pick my gloves until ten minutes before the fight. . . . I was cold.

(Beattie 2002: 233)

Psychologists often like to talk about attributional style, especially for events
characterised by success or failure, and that is why this is so interesting. Naz
wasn't used to dealing with failure, only success, and success was always internalised in his attributional reasoning. And not just successful at one thing (in
his mind), but Naz thought that he could do everything brilliantly, inside and
outside the ring, including his business dealings and driving his flash cars around
Sheffield (later he had a serious accident in one of these cars, which resulted
in a prison sentence). He thought that it was always going to be this way. And
when failure did come knocking, he made exactly the opposite kind of attribution – external (the bandages, the gloves, he was cold), restricted to this one
event and temporary. He had been doing this for years.

Naz had lost several times as an amateur, but these defeats back then had not
dented his confidence in the normal way. He never had any doubts. When I
asked him about the defeats in his amateur career in 1998, he told me

> I lost about five or six times as an amateur, but I could definitely say to
> myself that I have never walked out of that ring a loser when I'd lost. I
> always knew in my heart that I'd really won. I'd just lost for the simple fact
> that the judges didn't like my style and they'd gone against me. . . . I never
> *really* lost a fight as an amateur.
>
> (Beattie 1998b: 65)

Even though he said that he didn't want to make excuses, he did exactly that,
the equipment, the judges, the last-minute preparations, and these excuses
resided in an attributional pattern that characterised how he thought about the
world and his place in it. Brendan had worked on training him to think like this
over many, many years. You can get rid of doubt by blocking your thinking and
disguising your deeper reading of emotions and situation, but this may come
at a cost. Arrogance is always based on some sort of self-serving attributional
platform. But what was also interesting was there seemed to be no strong negative emotion to defeat, no shame. Maybe if you have an external attribution for
failure the intensity of the emotional response, when it does eventually come,
is never going to be as keenly felt.

But perhaps there is something in addition working here. In defeat people
expect you to show certain emotions, to be down, to be embarrassed, ashamed.
That's what they expect. Naz turned this into another competition, an alternative competition – one that he could still win. This was Naz talking about a
defeat in his fourth fight when he was eleven years old.

> My opponent never won properly; the judges simply went against me. I knew
> inside that I'd won. When they looked at a guy like me who had lost in the

ring but had a big grin on his face, they would look at me all funny. 'Why is he actually smiling when he's lost? Is there any way we can get this guy down?' This is what they were thinking. I'd be looking at them, knowing what they were thinking, and I'd be smiling at them. When I got back to the changing rooms I was still smiling and laughing because I knew that I'd won. It didn't really affect me in any way, but unfair judges; decisions did dishearten a lot of kids, and a lot of kids retired because they were getting robbed.

(Beattie 1998b: 66)

At eleven, he was already wearing a 'wicked gown' – 'gold, flashy with enormous shoulder pads' – and doing the Ali shuffle as soon as he entered the ring. He had been trained by Brendan for four years by then; his attributional style already firmly in place.

There is so much importance in the fear of failure literature on the role of emotion in directing our cognitions, but it works both ways – our thinking influences the intensity of our emotional responses. Brendan had shaped his emotional landscape by teaching him to internalise all success and deflect everything else. Naz had been the keenest student. In my interview with Naz in his own gym he referred to Brendan 'brainwashing' him. He said, 'He used to take me for walks every Sunday, and he used these brainwashing techniques – he was very, very good.' But then he added,

I'm going to be honest with you. I learned a lot from the gym, and I learned a lot from Brendan, and it was a fantastic step in my life, and it was a fantastic journey that I took, and I'll never forget it, never. It was an experience.

The father and son relationship between Brendan and Naz broke down in the late 1990s and they never spoke again. The fall-out was over money, as it always seems to be in boxing. But according to the author Nick Pitt (1998), there was something else – an argument brought on by Naz teasing Brendan once too often. Naz is reported to have taunted Brendan, with, 'What did you win, Brendan? Nothing. You never even won an area title.' Brendan apparently reacted to this, and then Naz administered the coup de grace with,

You know your trouble, Brendan? You never stood up to anybody. You never stood up to anybody in your life. You always let people bully you. Like the time with Mickey Duff [a boxing promoter] when he slagged you off and you just stood for it.

According to Pitt, 'Brendan became obsessed by what Naseem had said, as if Naseem had stained his whole life.' Pitt says that the comments stung so much because there was an element of truth in them,

But the person he had failed to stand up to was not Micky Duff, or indeed any of the others who had given him trouble over the years. It was Naseem himself. Brendan had allowed himself to be bullied by the bully he had, in part created.

Naz called Brenda 'Judas' for discussing their finances and these other issues with Pitt. Even as a practising Muslim, Naz would have known how much that would have hurt the Roman Catholic Ingle. In 2015, when he was about to be inducted into the International Boxing Hall of Fame in Canastota, New York, Naz said that he hoped for a reconciliation. Hamed said that Ingle too should be in the Hall of Fame with him. 'He's produced so many world champions. The time I had with Brendan was an amazing time. It was priceless,' he said at the time (*The Star*, Tuesday 16th June 2015). But there was to be no reconciliation. Brendan died in May 2018. Naz did not attend the funeral. He never doubted he did the right thing by not going, perhaps his arrogance and his way of thinking from those years of training prevents him from doubting anything he does. But perhaps all that comes at a great personal and psychological cost.

- Doubt and self-doubt are a central aspect of all sports centring on fear of failure.
- In boxing, doubt can be particularly acute.
- You can lose your self-worth in that ring, you can lose your livelihood, your friends, your routines, your support network, your self-respect.
- You can even lose your life.
- Those athletes across sport with the greatest fear of failure display greater anxiety levels.
- For individuals high in fear of failure, achievement events are threatening, judgment-oriented experiences that put one's entire concept of self on the line.
- Those with high fear of failure often adopt specific avoidance-based goals and 'self-handicapping strategies' that have negative effects on performance and psychological well-being.
- This is the challenge faced by the boxing trainer. He must stop that skinny kid losing the fight before he even steps into the ring.
- There is evidence that fear of failure is passed down through the generations.
- Fear of failure affects how the parents view failure and their own failures, and therefore how they respond to the failures of their children.
- The parents self-handicap in their own lives, and this is passed on to their children.
- Better not to try than try and fail and suffer all that shame and embarrassment.

- This is what goes on in some people's minds and is the mechanism of transmission.
- This trainer in this most humble part of Sheffield must break this chain of transmission.
- Families pass on their fear of failure to their children – Wincobank in Sheffield was full of families who no longer tried; Brendan Ingle had to block this process.
- I outline how he did it when he made a world champion, removing all doubt from him in the process.
- I also detail the consequences.

8 Manufacturing doubt

I grew up with smoking; it just seemed to be life back then in the 1960s and '70s. It was all around me; my clothes stank of it. Both of my parents smoked and on Saturday nights, the smoke hung like a low cloud in our front room. My father would be down in Paddy's Bar at the bottom of Barginnis Street, and my mother, my Aunt Agnes and a few close friends sat with their sherry, port and vodka in our front room and had a wee fag. The public bar was for the men in those days; the women and their wee drink were largely confined to the house on a Saturday night. They would sip at their drinks in the front room in front of the telly with the sound turned down, 77 *Sunset Strip*, and silenced images of glamorous Los Angeles, and they would pass the fags round, and maybe dream a little of Stu Bailey and Jeff Spencer, the private investigators from the show.

I would sit at the top of the stairs, waiting for my father to come home, needing the toilet out the back of the yard. I would spoil the glamour and the dreaming. 'I need to go out to the yard,' I would rhyme, that was how it was described, as rhyming, and probably how it sounded, until my mother eventually gave in. 'Out you go, then straight back to bed.'

You could feel yourself walking down into this thick cloud of smoke with sharp, visible boundaries that you could trace with your finger, halfway down the stairs. It was like walking into a cloud of acrid gas. The glasses would be pushed behind the settee and the fags held out of sight, as if there was some shame in it on their part, as I walked past, fanning the smoke away from me. 'Stop breathing it in, if you don't like it,' my mother would say. 'You can stop breathing for a few minutes, if it upsets you this much.'

One of the few photographs I have of my father is of him, down on his hunches, by our front door, feeding the pigeons in the street with a cigarette hanging out of the side of his mouth. He wasn't a big smoker compared to most in my street. My mother liked Humphrey Bogart in *Casablanca* ('Humph', she called him in a familiar sort of way, and her eyes would look all dreamy) and she loved how Humph or Bogie would talk with a cigarette in his mouth. She liked my dad with a cigarette hanging out of the side of his mouth; that is what I guessed anyway, as if he wasn't bothered either. Sometimes, I think my father was just trying to look cool and sophisticated for her. He died suddenly at fifty-one when I was thirteen. This sudden loss left

DOI: 10.4324/9781003282051-8

me feeling numb. I was angry and hopelessly lost, but it was an unarticulated anger about that great chasm of death. My Uncle Terence said that he had never seen a boy so close to his father, but after my father died, Terence said that I just pulled the shutters down and never mentioned him. I couldn't. It would have been far too painful. One night, however, I plucked up the courage and suggested to my mother that maybe my dad shouldn't have smoked and maybe, just maybe, he should have exercised more. However, my mother, inconsolable in her grief, could not bear to listen to any of this. She told me that I didn't know what I was talking about, that he had had rheumatic fever when he was a boy, that he had a heart murmur, and that my compulsive running, every day, which I started that same year he died, would do me far more harm than the odd relaxing fag. 'All that bloody sweating, do you think that's doing you any good?' She was crying as she said it: 'that sweating will give *you* rheumatic fever.'

Smoking always seemed 'natural' up our way, built into the very rhythms and patterns of life itself. You smoked when you were laid off from a job and when you were on the bus home at the end of a hard day in the shipyard or the mill, but only on the upstairs deck. You smoked when you had a lot on and when you had nothing on. The fags would be passed around when your friends popped in as a way of starting a social exchange. In houses, there would be an ashtray beside the settee, another beside the cooker, one beside the bed. They had to be handy, and the rooms were organised around smoking. In some of my friends' houses there were bottle-thick glass ashtrays, the thick glass stained by smoke, the ashtrays probably lifted from bars at the end of the evening ('they'll never miss them'). Some ashtrays were even more obvious in terms of place of origin – coloured, dented ashtrays with 'Guinness' in bold black letters on the side or 'Harp' in an insipid green. However, these ashtrays had little class and signalled none of the sophistication of smoking, or none of the glamour of *77 Sunset Strip*, which my mother and her friends desired. They all wanted to be 'the fella in the big picture' – that was the expression then, or at least his female companion.

My mother worked for a while in the 1960s for Ulster Plastics, and we had fancy chrome and transparent red plastic ashtrays where you pushed the cigarette through a lid into a smouldering cauldron underneath. The burning ash looked even more volcanic because of the thick red hardened plastic that gave off a Vulcan glow, and then the light chrome lid would jump back into position after you had pushed the butt through, with a little sharp yell, starving the cigarette of oxygen (eventually). It was like a cross between a child's toy and a utensil. I would ask if I could put my mother's cigarette out ('mind your fingers, that lid will cut them off'), and wished that I were old enough to have my own cigarettes to extinguish in this way.

The women from our neighbourhood did a lot of their smoking in their houses, certainly not on the street or on the bus ('too common'), and the fancy ashtrays would come out then. There were plenty to go around; Ulster Plastics never missed them. It was another part of the ritual of that most ritualistic of all habits. 'Pass me one of those lovely ashtrays, Eileen. I bet they were very expensive.'

My friends all smoked, starting at eleven or twelve, at the very beginning of secondary school. It was a rite of passage, like kicking the doors of the old pensioners from Legann Street (but only the ones that could still run after you like Charlie Chuck) and hoofing it back to our street. Or jumping nine feet off that overhang up Harmony Hill into the soggy ground at the other end of the wee millstream, without getting your feet too wet, or breaking your leg. 'Have you done it?' they would ask. 'Have you done the big Harmony Hill jump?' 'Have you had your first fag?'

I never liked smoking, but I can recall that first taste in a wee damp hut at the top of our street, egged on by an old night watchman sitting in the hut, there to watch over the new tar going down on the road. Duck, Jackie and Kingo sat huddled around as the old man's stained butt was passed around these young lads sitting on the wooden stools in the light of a paraffin lamp. 'It'll keep out the cold,' he said. 'Suck it right up into your lungs; it'll warm you up. It'll give you a lovely glow.' It was my turn to hold the damp little butt and try to suck it in. I was revolted at the taste, and the wet spit on the end of it, and they were all laughing at me, egged on by the night watchman with the stained yellow teeth.

My friends, too young to buy their own, would ask the bigger boys for their butts, or pinch a fag from the open pack of their parents on the kitchen table. That night it was a wee butt, just about to be discarded, the fingers of these young boys pinched around the stem already brown, already nicotine stained. It looked and tasted disgusting, and whatever was drawing them to it was not enough for me to overcome that most basic of emotions.

My mother smoked fags with filter tips. They were more ladylike in the sense that you didn't have to smoke them right to the end, and hold them in that miserly, manly pinch with hollow sucking cheeks. She always described smoking in a particular way: 'I'm just going to have a wee fag'. It was always 'a wee fag', as well as 'a wee drink'. Just as long as you put the diminutive in front of it, it never seemed that serious or harmful. But I didn't like the smell of smoke and I would fan the smoke away with my hand with these short, sharp little repetitive movements even when the house was full of her friends there for their wee fag and their wee drink on a Saturday night. My mother would tell me off for it. 'Our Geoffrey is at it again, always making a fuss about nothing. It's not going to kill you, for God's sake,' she would say. 'It's only a wee bit of smoke.' However, I was sure that it made me cough, it irritated my chest, but it was hard to complain without starting an argument. 'It doesn't do you any harm,' she would say. 'Or they'd ban it like everything else.'

On my lunch breaks from primary school, I would be sent down to the shops for my Aunt Agnes when she was on her break from the mill in Ligoniel. She worked in the carding room in Ewarts where the flax was combed and prepared for spinning and the dust from the flax hangs in the air. 'You'd have something to complain about if you had to work there,' my mother would say. 'That would make you cough.' My Aunt Agnes was a big smoker. It was always the same order every lunchtime, a half-pound of steak for grilling for my Uncle Terence (who was always on a diet) and a bone for our dog, Spot, who should have been on a diet but wasn't and then it was down to the sweetie shop for forty Woodbine for my Aunt Agnes ('and get an extra packet when you're down there, and you can keep the change'). In later years, it was Embassy for the cigarette coupons, which she collected, but the coupons were always collected for other people, and mainly for me. My Aunt Agnes would say that she smoked forty a day, but it was probably sixty or even more (and the third packet would mysteriously come out late at nights). 'Who's counting?' my mother would say, and my aunt had a very loud and very distinctive chesty cough that bent her double and lasted for an eternity, but this was never attributed to the cigarettes themselves, it was always attributed to the carding room and the mill. 'All the girls from the carding room have a cough like that for goodness' sake,' my mother would say, 'the dust gets into your lungs. Many people say that smoking clears the lungs, especially the menthol ones. Apparently, they are very good for the air waves.' My Uncle Terence hardly smoked at all, and Agnes had to hide the extent of her smoking from him. She didn't bother with the menthol ones in the end.

My Aunt Agnes was my mother's sister; my Uncle Terence was my father's best friend. They were more than an aunt and uncle though; they were more like a second set of parents. They lived at the top of Ligoniel in Belfast, we lived at the bottom. I would be allowed to go to their house on a Saturday night and sleep between the two of them. Her cough would keep me awake. All these years later, I can close my eyes and see her bent double, and I can hear the same hacking sounds.

The carding room, where my aunt worked, was one of the dirtiest jobs in the mill; my mother worked in the twisting room – one of the cleaner jobs. My uncle always said that my mother was a great one for the style, with her clothes and tipped cigarettes. She worked in the twisting room and looked like a film star, 'the best-looking girl in Ligoniel', my father always told me. I am now sitting years later reading some British Parliamentary Papers about the conditions of work in the flax mills of Belfast on the health of the workers. The report mentions the professional opinion of a Dr St George, surgeon to the Antrim Infirmary:

> I find that a large number of hacklers and roughers among males, and carders among females, suffer from phthisis, bronchitis, and asthma due to the irritation causes by the particles of flax being carried into the lungs and causing mechanical irritation.

The hot, damp condition of some rooms in the mills killed many prematurely, but in the carding room it was the dust. James Connolly, the Irish socialist leader, later famously executed in a chair by the British for his part in the Easter Rising, wrote a powerful manifesto to the 'Linen Slaves of Belfast' in 1913, condemning the 'condition of the sweated women of all classes of labour in Belfast'. There were many folk tales amongst the workers as to what would help these sweated women, including the use of whisky 'to clear the pipes', to allow you to breathe, but smoking is still to me at least one of the less probable antidotes, especially in this more modern era.

'Your mother was always fond of style,' my uncle would say. 'She could go to work dressed – your aunt couldn't.' She would come to our house every day for lunch: it was always the same thing, banana sandwiches and a fag. Soon after my father died, my aunt and uncle moved to Bath, just before the Troubles started. My uncle worked for the Royal Naval Stores, and he got a transfer because he was a Roman Catholic and he felt that he was being discriminated against in terms of promotion. I would spend all my long school holidays with them. My aunt was fifty-three and looked out of place in Bath among the foreign students, French schoolchildren, and middle-class matrons who invaded the Pump Room. While my uncle worked, she would wander about Bath in their first year there – the shops mainly, Woolworths and Littlewoods. She didn't think much of the Georgian terraces and, anyway, most of them were at the top of a hill. All those Woodbine and Embassy for the coupons made that hill very difficult to climb. My aunt got a job in the Naval Stores where my uncle worked, after twenty-eight years of inhaling flax dust. 'She couldn't believe her new job,' my uncle said. 'You could go to the toilet whenever you wanted to – and have a fag if you wanted one. If you'd done that in the mill, you'd have been out on your arse.'

They only came home once for a holiday, I think that it was too painful for them, they were far too homesick. However, there were not many holidays left. My aunt's health deteriorated. For two years, she couldn't walk more than a few yards at a time. When she inhaled, her chest made a loud whistle. The carding room, and all that saving for the Embassy gifts, always given to others, had taken their toll. She died in Bath, cardiac arrest, respiratory failure, pneumonia; the grim reaper had made no mistake with this one. But at the time of her death, one strange thing had occurred. My aunt and uncle had come to resemble each other quite closely. They had both gone grey, and old age had bent and short-ened my uncle and made him thinner, in a way that the steak for grilling never could; infirmity had made my aunt fatter – her body had filled with fluid, her abdomen had expanded. She had the same skinny legs but massive torso. She resembled a house sparrow at the time of her death.

After the service, we went to the pub and my uncle told anybody who was prepared to listen that he'd 'just buried his wife today', even though she had just been cremated. However, the smoking was never really blamed for this life foreshortened and sad in the end. I always thought that a little odd, even at the time. My uncle and aunt had many arguments about smoking, but it was never about the health aspects, more about wasted money and will power. 'It's not money wasted,' my aunt would say. 'What about the wee cigarette coupons?' Sometimes the arguments were about 'giving it up for a while to show who's in charge', arguments about human weakness. However, there was always that strong and resolute defence. 'It's my only pleasure.' It was that argument over and over again, 'it's not that bad for you they say; if I stopped now it would probably kill me.' Years later this was my mother's refrain, and then in the end she packed them in 'and I was never as unwell when I did.'

This, of course, is just a story of an ordinary family and smoking. Smoking was a part of life, and there must be millions of stories like this. But why was such a harmful activity so deeply ingrained in everyday life, so accepted, so uncontested? It's not as if the scientific evidence didn't exist, even back then. Dr Leroy E. Burney, the Surgeon General of the US Public Health Service had warned the public back in 1959 that 'The weight of evidence at present implicates smoking as the principal . . . factor in the increased incidence of lung cancer' (Burney 1959). However, the public paid very little notice. Boyd and Levy (1963) reported that in the United States in 1963, 78% of American men smoked, with the proportion of women smoking significantly increasing. They also reported that deaths from lung cancer in the United States had increased 600% since 1935, paralleling the rise in smoking. They expressed considerable surprise at the fact that the public were clearly failing to see the connection between cigarette smoking and lung cancer. Indeed, the American Cancer Society had reported that only 16% of the American public thought that the two things were related. Boyd and Levy say that

> This may be selective perception at work; that is, because the facts of the situation [smoking] are totally unpleasant and affect a basic habit, the mind rejects the message or, in effect, never receives it. Domestic sales of cigarettes in 1963 hit a record of $512 billion.

They also point out that there were decreases in cigarette smoking in only two years since 1935, and they were in 1953 and 1954, which happened to be the years immediately after the negative publicity emanating from the first set of studies suggesting a link with cancer. However, from 1954 onwards the figures started to rise again 'attributed in part to filters, which now account for over 50% of all cigarette sales'. For Boyd and Levy, this change was not attributable to the 'technical' aspects of the filter tip but the psychological aspects. They say that it allowed smokers, particularly female smokers, 'to rationalise the habit', to feel safe. When I watched my mother and my aunt refuse non-tipped Embassy, I knew exactly what they meant. The cigarette companies guessed that women are the more cautious sex.

Boyd and Levy wanted to address the question of how we can prevent young people starting to smoke in the first place. In the early 1960s, the evidence was that children were starting to smoke at younger and younger ages, with boys starting younger than girls. Boyd and Levy suggested that we must think beyond the basic facts that smoking is a deeply ingrained habit rooted in biological addiction. These are critical factors, but they are only a small part of the story. They argued that we need to better understand the psychology of smoking, and here they turned their attention to that guru of the smoking industry Dr Ernest Dichter. Dichter had been the president of the Institute for Motivational Research, who had conducted the pioneering research into the 'strong psychological forces' that operate with smoking back in the 1940s. He had carried out his original groundbreaking research not to stop people smoking but to help them get addicted. Given how effective Dichter had been, they suggested that this might well be a good place to start. 'How else could the habit continue to exist?' they ask, 'when so many smokers classify it as being unhealthy, wasteful, dirty and immoral? Certainly, powerful motivations must be operating to sustain cigarette smoking in the face of these negative attitudes.' Powerful motivations indeed, not just assessed, but manipulated by Dr Dichter and other psychologists using *depth* techniques to probe the human mind and reveal its unconscious motivations and then direct them in certain ways.

Ernest Dichter was by training a psychoanalyst, another of that wave of Jewish refugees from Nazi Germany arriving in the United States in the late 1930s. He was obviously taken by the entrepreneurial spirit of his adopted country and he had a simple idea that was to have a profound impact on all of our lives. Psychoanalysis was providing new insights into how the mind works; the emerging (and *now* somewhat clichéd) metaphor was that the human mind is like a glacier; most of it is hidden from view. Human beings may be rational creatures, but not rational all the time, and there are unconscious forces that govern much of our lives. Could marketing tap into these unconscious forces? Dichter decided to turn his attention away from the curing of neuroses using psychoanalytic methods to the application of psychoanalytic understanding to marketing. There were more similarities than differences between the two enterprises, he thought. People say that they are not subject to neurotic complexes, and people say that they are not open to manipulation through advertising. However, people say many things; the reality might be quite different. Both enterprises need a clear model of the mind. You need to see below the surface, to *access* the unconscious to cure neurosis and to *manipulate* the unconscious to sell brands. If you want to understand how people think and feel about products, you cannot just ask them directly. Psychoanalysts would never do that; it would be ludicrous ('please tell me about your neurosis'); you need more indirect and more intense methods. You need the psychoanalyst's couch, but on an industrial scale.

His first commercial project was on the marketing of milk. He suggested using more indirect methods for his market research. He interviewed consumers without ever asking them to talk directly about milk. This approach did not impress his employer ('Why is he asking them *that*?'). However, this did not dent either his self-belief or his belief in the truths that psychoanalysis had highlighted. He wrote to six of the biggest American marketing companies, introducing himself as 'a young psychologist from Vienna' with 'some interesting ideas which can help you be more successful, effective, sell more and communicate better'. He emphasised his connection with Freud, whose views on the importance of the unconscious were at that time becoming more familiar through the popular press in the United States (although even then often parodied in the press). Dichter had been trained in psychoanalysis in Vienna but not by Freud himself. However, he had lived on the same street as Freud in Vienna and he liked to weave this into the conversation whenever he could. Dichter liked to point out a few simple facts from psychoanalysis to the marketers of the day. People do not know themselves well, so there is really no point in asking them, he would say. You cannot ask people why they are neurotic; they have no idea, he would say this over, and over, again. So much of life is unconscious. Similarly, he argued, there is no point in asking people why they choose one brand over another. It is all about unconscious associations and subconscious impulses, repressed desires, defence mechanisms and guilt. That is the stuff of neurosis and the stuff of everyday desire, including the desire for brands. Freud had focussed on one set of applications of psychoanalysis; Dichter was going to focus on a different set. That was his mission; that was what unconsciously drove *him*.

It is important to set all of this within its natural context. A much simpler understanding of the human mind drove market research at the time. The traditional view was that people could report why they preferred some products rather than others. They knew what they liked about certain products and what they didn't. The big market research companies commissioned hundreds of surveys where people were simply asked why they bought one product rather than another. Vance Packard, in his classic 1957 book *The Hidden Persuaders*, refers to these methods as 'nose counting'. The problem, according to Packard, was that what people said retrospectively, and in the context of wanting to be helpful to the interviewer, often bore very little relationship to what they actually did. Packard says that informants wanted to appear to the world as 'sensible, intelligent, rational beings'. But how rational are people *really* in their decision-making? That was the big question. Dichter said he already knew the answer to this based on his psychoanalytic training; it really was very simple. 'Not very', was his answer. Moreover, some new research coming out of the Color Research Institute of America at the time was backing him up on this.

The Color Research Institute of America was another new development in rethinking how the human mind works and applying it to the commercial world. It was founded in the 1930s by Louis Cheskin, another clinically trained psychologist. Cheskin is identified as the other main contender to be considered as the originator of the 'depth' approach to marketing in the 1930s. He also argued that consumers make unconscious assessments of products and that these are not just based on the product itself but derive from all of the associated characteristics, including the sensory input from the product, all processed automatically and unconsciously (Cheskin 1951). One major sensory feature is the colour of a product, which is rich in both biological meaning (consider the colour 'red', for example, and its unconscious associations in the natural world – lips, blood, sex, etc.) and in symbolic meaning (where 'red' is often associated, again unconsciously, with festivity, danger and, of course, sex again). These unconscious sensory impressions from the product (or its packaging), Cheskin argued, can transfer directly onto our perceptions of the product itself, including its perceived value, price and quality. It also affects our emotional response to the product. In one market research study, the Color Research Institute tested package designs for a new detergent and housewives tried out three 'different' detergents in three different boxes (which were either yellow, blue, or 'blue with a splash of yellow') on their weekly wash. The verdict was that the detergent in the yellow box was too harsh for their clothes – 'it ruined them,' many of the respondents complained – whereas the detergent in the blue box was not strong enough and left the clothes still dirty. The detergent in the blue box with splashes of yellow was 'just right', they said. 'It cleans my clothes well but doesn't ruin them.' The detergent was identical in all three boxes. The colour of the packaging affected the respondents' perceptions of the effectiveness of the product. Unconscious associations, manipulated by the marketer, could determine our preferences.

Cheskin concluded that asking consumers about products or what influence their preferences is not a very informative way to understand the processes involved. It is what people do rather than what they say that matters, and we should not underestimate the role of the unconscious in this. Cheskin's company advocated putting Del Monte peaches in a glass jar rather than a can ('it unconsciously reminded them of their grandmother bottling fruit'). His company suggested putting a sprig of parsley on tinned Spam to signify 'freshness'. It was the research of this company that found that when you changed the colour of the 7-Up can, with a 15% increase in the amount of yellow on the can, it led to serious complaints about a change in flavour. Too 'lemony', consumers protested about the drink whose flavour, of course, had not been touched. It now tasted 'lemony' because consumers had been unconsciously primed with that lemon association through the colour of the can. This research was starting to question the more rational model of consumers, and how the human mind works.

Ernest Dichter was now at the forefront of this, and he developed a completely new *approach* to uncover the more irrational and unconscious side of consumer behaviour. Dichter said that we need to start at the very beginning, indeed with our first point of contact with the consumer. Interviews can be revealing for market research, he argued, but these interviews need to fundamentally change. He suggested that much smaller numbers of respondents should be interviewed (after all, many of the great insights of psychoanalysis were based on single case studies). But these interviews needed to be much longer and much more in depth, on the assumption that if you let people talk long enough, you may be able to find something interesting in the associations between concepts that come tumbling out in spontaneous speech often in unguarded moments when the ego is less in control. You also need to get respondents to talk indirectly (rather than directly) about the product. You need to get them to verbalise their feelings, you need to listen carefully, you need to check and crosscheck, and you need to listen for inconsistencies. You do not take what they say at face value, you need to understand instead the symbolic importance of products in people's lives, and you need to interpret what they say. You need to look out for defence mechanisms and projection. His approach clearly owed a lot to some of the basic processes of psychoanalysis, and it relied heavily on a number of its core concepts.

In a retrospective account of the first part of his career, Dichter outlined the principles underpinning his approach to understanding and measuring human motivations (Dichter 1960). His explicit goal in publishing this retrospective book in 1960 (and this is an interesting admission on his part) was 'to set the record straight'. He thought that there many misunderstandings had developed about how 'shrinks' like himself were manipulating the American public. 'I think the time has come for a little factual and unemotional clarification,' he wrote (Dichter 1960: 33). In an attempt to do this, he said that he needed to explain his fundamental understanding of the basis of human motivation.

He argues that 'most human actions are the result of tensions. Whenever tension differentials become strong enough, they lead to action' (Dichter 1960: 38). The example he uses here is buying a new car. The typical 'tensions' in a family derive from one's children complaining about the old model, their friends' dads having better cars, ads that tell you how good other cars are, your old car breaking down and so on. He says,

> We are dealing with a series of events, some coming from the inside, some from the outside, some from technical factors, others from psychological factors, all building up to a tension which results in action. The tension differential . . . has become so great that the action is finally triggered off.
> (Dichter 1960: 38)

And notice the way that he stresses the role of the children in the family as being important, indeed critical elements in producing this 'tension' and thereby acting as instruments of change.

Underpinning his approach, he says, there are three very basic principles. The first is the 'functional principle'. He argues that we cannot explain why somebody buys the same make of car each time or chooses a particular brand of cigarettes unless we know why they buy cars or smoke in the first place. We need to understand car buying and smoking in their natural context. For example, he says that in the case of smoking we need to 'analyze smoking in a way that will not disturb its natural ties with all its related activities and phenomena such as working habits, leisure time, occupation, health etc.' (Dichter 1960: 39). He says that there is nothing new in this approach; this is really just applied cultural anthropology (rather than primarily psychoanalysis in action), and he quotes Margaret Mead, who wrote, 'Anthropological science gains its importance from the fact that it attempts to see each individual which shares a common culture in its natural context' (cited by Dichter 1960: 39).

The second core principle that Dichter identifies is the dynamic principle. He argues that human motivations change during a lifetime as a function of aspiration. Motivation and consumer choice are influenced by previous experience and where the individuals concerned are hoping to end up. In other words, one needs to take a longitudinal perspective on the nature of the behaviour in question. A major focus for Dichter was the first experience of any product – the first car (always the most important symbolically), the first cigarette (for example, mine in that watchman's hut), the first fur coat (these were, after all, different decades). He would get his participants to talk at length about these first experiences in order to understand their current motivations and how these motivations had changed with time. I laugh when I think of what he might have made of my first experience of smoking, sucking on the night watchman's old dirty cigarette butt and feeling nothing but revulsion. At least it put me off smoking; I am thankful for that. What other associations might *he* have seen?

The third of Dichter's core principles was 'the principle of fundamental insights'. He says that the point about human motivation is that we, as actors, have no real idea why we do one thing rather than another.

> In practising research on human motivations, we feel it to be our duty to get down to fundamental insights, to accept the fact without fear or embarrassment that quite a number of human motivations are irrational, unconscious, unknown to the people themselves. This principle means that most human actions have deeper motivations than those which appear on the surface, motivations which can be uncovered if the right approach is used.
>
> (Dichter 1960: 45)

Of course, if you ask people why they do something, they will give you an answer ('we are not allowed by our culture to admit true irrationality as an explanation of our behaviour'), so he avoided asking them. Instead, he allowed them to talk indirectly and in depth ('tell me about your first cigarette'). He encouraged them to disclose their emotions and to talk about the extremes ('tell me about the best beer you ever had and the worst beer you ever had'). The reasoning here was to mobilise 'true feelings' and 'real experience' and to move away from 'considered opinions'. He encouraged them to be specific. When people talk in terms of generalities, it is easier for them to present a 'rational' and considered view of things. However, perhaps most importantly of all, he encouraged them to be spontaneous and then he analysed them carefully to work out their true feelings and intentions. In other words, this was an approach that did not take what people said at face value.

> Many of the aspects of depth interviewing are borrowed from the approaches used in psychiatry, where the problem often is to understand the real reason for a person's behaviour. We employ these techniques continuously in our daily lives. When the hostess keeps urging us to stay a little longer, but yawns at the same time, most of us don't need any knowledge of depth interviewing or of psychology to detect a discrepancy between her statement and her actual feelings. We leave.
>
> (Dichter 1960: 285)

He filmed his respondents interacting with various products (women were often surprised by how much time they spent sliding their fingers over a bar of soap to test its smoothness). He used psychodrama, where they acted out their relationship with the product (the sound of the fist on the glove is a crucial dimension for baseball gloves). He used various projective techniques ('imagine that you are a little boy looking though a keyhole into a kitchen, ten years from now. What do you see?').

His first major professional success was with the Compton Advertising Agency to promote Ivory soap, whose sales had significantly slumped at that time. 'The soap that floats' had been discovered by accident in 1879; it had been performing extremely well for many years. The standard market research approach had been to ask consumers why they chose this particular product, or why they didn't. Dichter wanted to apply his 'functional principle'. He said that there was no point in trying to promote a brand of soap before you understood more about the psychology of bathing, so he began by interviewing 100 people at various YMCAs around the country, in his usual non-directive way. 'I decided to talk to people about such things as daily baths and showers, rather than to ask people various questions about why they used or did not use Ivory Soap' (Dichter 1960: 33). Bathing, he discovered, had all kinds of hidden psychological significance. It was not just to do with washing dirt away for some people; rather it was a process of psychological cleansing as well. As he himself put it, 'You cleanse yourself not only of dirt but of guilt.' The slogan that he came up with was 'Be smart and get a fresh start with Ivory soap. . . . and wash all your troubles away.' 'Troubles' was used here to represent implicitly all of those guilty secrets that could be washed away with the right type of soap. It might sound implausible or far-fetched. How can soap wash away guilt? However, Ivory soap had a whole series of *pure white* connotations helping you to bathe, to look forward to getting dressed and going out, helping you to focus on what lies ahead rather than what lies behind. It represented a cognitive focus on the future rather than the past. If you can use ritual to shape your cognitive focus, then you can start to deal with the emotions that attach to thinking too much, about what you have done, like guilt. This all made some intuitive sense to Ernest Dichter. However, much more importantly, the campaign was a great success.

But bathing cannot just be about ridding oneself of guilt. That would be ridiculous. It would be like suggesting that a cigar *always* has some sort of sexual significance regardless of context. After all, did Freud not once say that a cigar is sometimes just a cigar (although interestingly although this quote is often attributed to Freud, it turns out that he himself may never have actually used these words [see Elms 2001], but the point still stands). Every instance of bathing could not possibly represent an attempt to deal with guilt (is there that much guilt in the world?), so Dichter also suggested that bathing sometimes has quite a different psychological function. He says that bathing is 'one of the few occasions when the puritanical American [is] allowed to caress himself or herself while applying soap'. It's not about getting rid of guilt any more, it is about enjoying possibly the sins of the flesh. If you are going to market soap, the importance of the guilt-free caress could also be very important, Dichter argued. That was another starting point for his thinking, and then his psychoanalytically trained marketing brain took over. There are very different kinds of caress, depending upon who is doing the caressing and who is being caressed. The touch, after all, is one of the most powerful and ambiguous forms of communication, which requires careful consideration (Beattie and Ellis 2017). The caresses of, for example, a mother or a lover are very different things, and Dichter argued this sort of distinction must be considered carefully in the marketing process. Some soaps (like Camay) could be constructed as sensual, the caress of an indulgent 'seductress'. For other soaps, like Ivory soap, the caress is maternal and caring. Thus, back in 1939 he was developing the concept of the 'personality' or the 'image' of the product. This is something that we take for granted now, but was very original at the time, groundbreaking even. Dichter argued that marketing campaigns could be built around these personalities of products, uncovered in the first instance through the depth interviews and then systematically developed in the campaigns themselves. The 'personalities' of some products, originally identified by Dichter, stayed in place for the decades that followed; they became the product that we all know and recognise today.

One Camay ad from the fifties shows a fresh-faced bride carried across the threshold by the young groom. She has bright red lipstick that draws the eyes. Her lips are slightly apart, exposing bright white teeth. The groom is holding her tight around the waist; she is pulling him towards her. The copy reads 'Your skin has a fresher, clearer look with your First Cake of Camay!' 'First Cake' is capitalised and underscored with a red line, drawing the eyes between the red lips and the red line. Camay is described as 'cake' (as some other soaps are), indulgent and to be enjoyed; it is sensual rather than functional. 'First' cake connotes the first time, unconsciously signalling that it is the first time for the young wholesome bride; in other words, she is a virgin. There are red roses around the packet of soap and the 'undressed' bar of soap in the image beneath the main picture. That is the third use of red in the ad. Of course, red roses communicate romance but at the same time, roses are associated with the woo-ing process. The man sends the woman red roses before a date, it is the will she or won't she phase, the wait. Tonight is the end of that long wait. The caress is sensual, like the soap itself.

Another ad from the same era shows a woman in a bath, bubbles cover her modesty. Again, she is fully made up, with bright red lipstick unblemished and expertly applied, and a mouth slightly open. Her hands luxuriantly stroke each other with a white bar of Camay caressed in between them. She looks dreamy, her eyes half-closed, but in the reflection in the mirror beside the bath her eyes look fully closed. She looks as if she has just been sexually satisfied. The copy reads 'you'll be a little lovelier each day with fabulous pink CAMAY.' However, there is more text on this one. There is little ambiguity about the message. 'When you surrender to this luxurious caress . . . feel Camay's gentle lather enfold you.' This is seduction and beyond, it is surrendering to the caress of a seducer, it is guilt-free sex with a bar of soap in the bath.

Ivory soap was not like that; it was never a seducer. It was pure white, and 'pure' was its main associative connection. Therefore, it needed to be marketed differently. It was more maternal and comforting; it was a different kind of caress. 'Wash your troubles away', the way a mother might comfort you (and forgive you for the things that you've done, thus dealing with feelings of guilt). The ads built on this maternal image. They often showed a baby with 'That Ivory Look' ('what is the first thing that comes into your mind when I say, "mother" . . . "baby"'). The copy reads 'Babies have That Ivory Look. . . . Why shouldn't you? . . . . doctors chose it ahead of all other soaps for today's complexion and for yours.' Another one reads 'Imagine this! Best costs less than all the rest.' Again, there is a picture of a baby. This will be the housewives' choice, doing the best for her family, careful with cost and family finance, taking advice from the family doctor, staying young to hang on to her husband, washing your anxieties about this away in the bath (regaining your baby-faced complexion with soap), allowing him to wash his guilt away. Two bars of soap whose prototypic 'personalities' are borne out of psychoanalytic theory – the wife as whore and the wife as mother. You could not ask consumers directly about these things; they would laugh in your face or punch you ('who are you calling a whore?').

Of course, you can easily criticise this type of approach, taking the unconscious and trying to manipulate it, but Dichter argued that there was hard data to evaluate the hypothesis. Not necessarily scientific data per se, of course, but the sales figures, in black and white, to see whether the idea was working or not ('better than any laboratory experiment', Dichter said). Sales of Ivory soap shot up. By 1979, according to *Advertising Age*, Ivory had sold more than 30 billion bars. I do recall that when my mother's old mill house was demolished in the 1980s and she got her first house with a bathroom in North Belfast, her very first purchase was a 'cake' of Camay soap. She deserved a little luxury and a little indulgence, she said, after all those years in an old condemned mill house. She had forty years of advertising firmly entrenched in her head. She knew the personality of Camay soap inside out.

Dichter now turned his attention to cars. The brief here was to understand why the new Chrysler Plymouth had not taken off in the way that had been anticipated by the company. The report from Chrysler's own marketing agency, J. Stirling Getchell, Inc., makes fascinating reading.

> Dr. Dichter proposed the use of a new psychological research technique to get beyond the limits of current statistical research in an understanding of the factors which influence the sale of cars. Quite frankly, we were at the onset as sceptical of the practicability and value of the proposed study as, we learned later, executives of the corporation had been when first approached.
>
> (Dichter 1960: 289)

Chrysler were interested in two main questions: why do most car buyers buy the same make of car as their previous vehicle (estimated to be around 70% at the time), and what influence do women have on the purchase of cars? The feeling amongst marketers at the time was that all previous answers to these two questions, based on the standard surveys, were not entirely satisfactory. The standard answer to the first question was that we tend to buy the same make of car out of 'habit' or 'loyalty'; marketers had no answer at all to the second question. Dichter was particularly scathing especially about any attempt to explain behaviour in terms of habit:

> They are similar to the type of pseudoscientific facts assembled over many decades by psychiatry which explained the fear of narrow places as being the result of claustrophobia. Translating this into simple language, the statement would read, 'He's afraid of narrow places because he has a fear of narrow places.'
>
> (Dichter 1960: 45)

Dichter used his depth interviews and offered new answers to these questions, answers of a type that they had never seen before. He said that cars (like soap) have personalities ('the more you live with, or experience your car, the more personality it has'). He also said, based on his psychoanalytic training of course, that cars clearly project our innermost fantasies. Convertibles, he said, symbolise freedom and they project the fantasy of being young, free and single again. For this reason, Dichter argued, wives will rarely allow their husband to buy a convertible. This has nothing to do with economy or appropriateness or even (in those sexist times) 'what a convertible might do to their hair', rather the women respond negatively to the symbolic significance of the convertible for their husbands (although they would never allow themselves to think such thoughts explicitly in those terms; such thoughts would be far too threatening). Women unconsciously and implicitly understand the danger that convertible pose to their marriage. Dichter reasoned, therefore, that car dealerships should put convertibles in the front windows of their showrooms to draw in the middle-aged men, racked by wish fulfilment, yearning to be young, free and single again. However, in addition to this, the car dealerships had to ensure that there were plenty of attractive sedans just behind the convertibles, so that the husband and wives (driven there by differing desires and motivations, explicit and implicit) could make harmonious joint (compromise) decisions. Women, he asserted, have a critical economic decision-making role in the family, and the final decision, one way or another, is always going to be a joint decision.

Dichter's overall view on the role of women in car buying was that 'Women influence car buying directly or indirectly in about 95 per cent of all car purchases' (Dichter 1960: 310). Once you recognise this fact, then there are certain logical consequences. According to Dichter, women are 'the economic conscience of the family of the average income bracket', and, therefore, you have to give them 'moral permission' for the 'sinful extravagances' of the new car. At the time (1939), many women still considered cars to be a luxury; the goal was to persuade them that they were a necessity.

Then Dichter turned his attention to the question of why most car buyers go for the same make repeatedly. 'It is not habit,' said Dr Dichter. 'That's too easy and it's not even a proper explanation; the real reasons are separation anxiety and fear.' He argued that we have a degree of psychological attachment to the old car and we suffer from 'separation anxiety' when we think about getting rid of it. According to Dichter, 'People do not buy new cars as much as they sell their old cars.' We cannot bear to part with the old car despite all the 'tensions' that have built up – 'The old car has become a part of our personality. To give it away is like giving part of our personality away.' He referred to this as a form of 'separation anxiety', but it is an odd sort of 'separation anxiety'; it is more like a form of splitting of the ego with a part of ourselves given away. So, Dichter argued, we essentially compromise, we buy a newer model, but the same make to decrease the separation anxiety. This is one major factor in sticking to the same model. Then there is the fear of embarrassment. Men feel 'more comfortable buying a car in which the basic elements and engineering features were familiar to them'. Men do not want to look foolish; they do not want to buy a car with which they are unfamiliar, unsure of how things work, uncertain in front of others.

This was all radical stuff, quite different to anything that had been offered up before to Chrysler, or indeed to any other manufacturer at the time. It was a new way of thinking. It was even more radical given that the whole Chrysler Plymouth campaign up to that point in 1939 had emphasised how different this car was. The advertising copy actually read 'This car is different from any other one you have ever tried.' According to Dichter, this campaign was counterproductive – it aggravated the fear of change; it made the whole thing much worse. Dichter suggested that they should base the campaign around reducing the fear, emphasise the fact that it would only take a few minutes to feel at home in the new car.

This campaign was a major success in commercial terms, and the J. Stirling Getchell, Inc. agency took him on as a full-time employee. In subsequent years, Dichter turned his attention to lipstick (a phallic shape has a massive effect on sales, according to Dichter because it offered a subconscious invitation to fellatio, 'but one has to be careful not to go overboard,' he warned, 'and make the parallels too obvious'). Then, perhaps most controversially of all, he turned his attention in the late 1940s to cigarette smoking (Dichter 1947) and here, one could argue, he made the biggest difference of all, and we are all, one way or another, still trying to deal with the consequences. His starting assumption was that cigarette advertising up to that point (like so much other advertising) had got it seriously wrong. The ads at that time were all designed to emphasise the flavour of the cigarette, or how mild they were (The Chesterfield ad from the forties had Alan Ladd say 'I like Chesterfields – they're my brand because they're MILD'; 'Lucky Strikes means Fine Tobacco'). Based on depth interviews with 350 smokers, Dichter concluded that things such as taste, mildness or flavour were 'minor considerations' when it comes to smoking; the main appeal of cigarettes was the range of psychological pleasures that you get from them. He argued that, from a psychological point of view, cigarettes work in several distinct ways. Firstly, they allow you to behave like a child again, able to 'follow your whims'. They offer a 'legitimate excuse for interrupting work and snatching a moment of pleasure'. Like children, we crave rewards – 'a cigarette is a reward that we can give ourselves as often as we wish.' Dichter argued that they should use this insight of self-reward as the basis for a marketing campaign.

However, that was only part of his observations about the nature of smoking as an *activity*. Some of his respondents had also commented that with cigarettes, you never really feel alone. Perhaps, this was tapping into the primitive concept of fire, a warm glow, a conditioned stimulus rooted in our evolutionary past associated with the group assembling around the fire. One of Dichter's respondents said when I 'see the glow in the dark, I am not alone any more', and Dichter added that the use of cigarettes to combat feelings of loneliness and isolation was critical. He also added that 'the companionable character of cigarettes is also reflected in the fact that they help us make friends.' This insight formed the basis of several marketing campaigns.

I have music in my head from my earliest days of childhood, and a tag line that I cannot forget. I can see the black-and-white TV in the corner of the room. The smartly dressed handsome man in the raincoat and trilby walks slowly through the wet and deserted London street. He is walking very slowly as if he has nowhere really to go. Why is he on how own at that time of night? Has he been stood up? He gets out his packet of cigarettes, lights up and smiles briefly. I close my eyes and I can see that contented smile of his. That is the hook. Then the words come over the music: 'You're never alone with a Strand.' That slogan will never leave me. I may forget the most important words that my children have ever said to me (and even shamefully their very *first* words), but I cannot forget those words, carefully selected by the mechanics of the mind. This 'lonely man' ad ran from 1959 onwards. 'Strand . . . the cigarette of the moment. Strand, the new tipped cigarette, wonderful value at three and tuppence for twenty.' Over the next four or five decades I saw many lonely people reaching for a fag to attempt to alleviate their crushing loneliness, primed by this ad playing at some point in their past.

The emphasis in his approach was on understanding the range of roles and functions that cigarettes can play in people's lives. In other words, he wanted to start with some psychological understanding of what cigarettes *do* in people's lives, following again what he had learned from his reading of Margaret Mead and other cultural anthropologists. His functional analysis built from there. Repeatedly he found that people smoke to relieve tension and as a reward for something that they have done. They smoke to reduce stress in anticipation of an event. They smoke as a symbolic statement about how daring they are. They smoke as a way of bonding with others. They smoke as part of a ritualised performance that requires little planning, but it allows for a projection of sophistication. Cigarettes are used before sex because people are nervous, and after sex as a relaxing reward. All of these observations are critical in the marketing process.

Then, of course, there is the oral pleasure that derives from smoking, 'as fundamental as sexuality and hunger'. Here, he was reminded of some great psychoanalytic truths about early stages of psychological development and frustration and how people deal with it. What is smoking in essence? It is a form of behaviour, heavily overladen with symbolic and social connotations, where you put something in your lips to comfort yourself at times of stress or frustration, or as a reward. It is about oral gratification, as infantile in the context of Freudian theory as sucking your thumb. However, it is a socially acceptable way of obtaining this oral gratification by simultaneously sending out a powerful message about virile maturity and potency. It is so powerful because it simultaneously satisfies infantile desires and yet symbolically sends out a very adult signal. Young people smoke to look adult (as my friends did when they were eleven or twelve), and old people smoke to look 'potent', according to Packard 1957).

Dichter also highlighted the power of lighting a cigarette (as well as, of course, its more 'social' connotations). It was the power of fire that helped define and shape *Homo sapiens* in evolutionary terms. So, we carried this fire with us in our evolutionary history and we also did so in our own personal history, and it represented a milestone in development. Indeed, you knew you were leaving childhood behind when you wanted to carry that one object. It was the object that you now most desired, not a toy any more, no more Fort Apaches or missile launchers, no more knights on horseback. It was a lighter, just a normal little tin lighter, but one that could never come in your Christmas stocking. It represented another break of that family bond and the building of that bond with your mates, your alternative family. I don't know which of my childhood gang got the first lighter, but I do remember this Prometheus passing it round and talking us through it, like it was a complex bit of machinery. Pull the little wheel back slowly, slowly now, now let it spark up. Feel the power. Not every time, sometimes it would take four or five strokes of your soft-skinned, boyish thumb until it fired into life. According to Dichter, this was a design feature of lighters, a case of variable ratio reinforcement (like slot machines) to keep you hooked. However, it was worth the wait, it was a genuine reinforcer; you had the power of fire in your hands. This was a primitive power rooted in group dynamics and survival, full of deep emotional significance that any evolutionary biologist might understand. 'Smokes, lads, light up,' and all the fags held lightly in the various damp childish mouths would gravitate towards the tip of the fire, all at more or less the same time, as if there was an invisible cord pulling them in close.

I *found* my first lighter, which was very fortunate; in other words, I didn't have to steal or buy one (which was always going to be unlikely given both the cost and the embarrassment). The lighter was plastic and a sort of orangey red, like cheap lipstick. Not one colour or the other, perhaps designed for girls who couldn't make up their mind about the colour of their lipstick. Maybe they wanted a lighter to match. So, on the back seat of the cinema, the lighter would draw attention to the lips. Only certain sorts of girls would light the fag of the man anyway, my mother always said. The lighter was very temperamental; it was on its own very variable ratio schedule. I carried it with me even though I didn't smoke. 'Anybody got a light?' One of my mates would say. And I would pull it out. 'Are you not having a fag? We can share if you don't have the readies to buy your own.' 'I'm alright,' I said.

Duck called them coffin nails, funny even then. 'Who wants a coffin nail?' And they'd all stick out their hand to the fella in the big picture, with a pack of 20. Nobody asked where the packet of fags came from. Most of the lads could just afford fags bought individually from the sweet shop ('one Park Drive please, Mrs B'). And they all sat in the park blowing smoke rings, sometimes straight up with their heads right back, and putting their lit fag through the fading halo, a little stabbing movement with a pencil-thin dart. It was all only play, I knew that. The playing of roles they had seen and they somehow aspired to. They blew smoke rings like cowboys from the Westerns, the bad 'uns who needed time to think, and stubbed their cigarettes out like gangsters with a job to pull. It looked business-like, serious, they looked like they were not to be messed with, time to leave. All our parents smoked, but they looked different when they smoked. We were recreating smoking for ourselves, the way every generation does, with new role models, and fashions, and all sorts of symbolic signalling of, and for, the moment. Our little gang looked cool smoking, or that's what most of them thought; our parents looked just sad, gasping at theirs.

I sat years later with Duck in a dark bar in Belfast in the middle of the day. He was still smoking, of course. We talked about old times and our boyhood scrapes. Duck was shaking with laughter as he told one story, until his fag ash dropped on the table. He saw me watching the drooping ash fall almost hesitantly to the smeared table top. 'You still not smoke?' he asked. 'You don't know what you're missing. It's one of life's little pleasures that they can't take away from us.' And he lit another one, almost immediately, with a shiny gold lighter. He saw me admiring it; at least that's how he read my look. 'Only the best, these days,' he said. 'Still the fella in the big picture,' I offered. 'Still the fella in the big picture,' he replied, and he drew the smoke deep into his lungs and held it there, as if it might never come out. He was saying to me that I am the man in control. However, I never ever agreed. And then he blew it my way, until I gave a brief and embarrassed cough.

Of course, Duck is dead now. We talk about lifestyle choices and Duck made his, but Ernest Dichter helped direct these choices. In order to promote smoking in Duck, or my mother, or Aunt Agnes, Dr Dichter says that we have to recognise the psychological needs associated with smoking and what smoking provides us with in our everyday lives, be it in uptown Manhattan or in the less salubrious surroundings of a politically divided city in Northern Ireland. At the same time, Dichter argued, we have to deal with the essential psychological conflict that smoking generates. He wrote, 'One of the main jobs of the advertiser in this conflict between pleasure and guilt is not so much to sell the product as to give moral permission to have fun without guilt.' There were clearly echoes of selling the Chrysler Plymouth to the wives of the car buyers here. Learning to smoke in childhood or in the teenage years is always going to be associated with a degree of guilt, no matter how defiantly it is initially carried out. The guilt associated with smoking was compounded by the fact that certainly by the 1950s there was growing evidence that smoking was indeed very harmful (although Dichter is cynically dismissive of the accumulating evidence – 'Scientific and medical studies on the physiological effects of smoking provide a confused picture: Some conclude that smoking is harmful; others deny it. This same confusion prevails among smokers themselves'). Cigarette companies were trying to use the message that they would not kill you as part of their pitch. This was often done by having doctors (and dentists, interestingly enough) recommend particular brands as being 'healthier'. Dichter thought that this was fundamentally misguided. It was unconsciously associating cigarette smoking with increased mortality; the 'not' was not necessarily the critical element in how such messages were interpreted by the public. Dichter concluded that smoking offers 'a psychological satisfaction sufficient to overcome health fears, to withstand moral censure, ridicule, or even the paradoxical weakness of "enslavement to habit"'. Ads now focussed on powerful men relaxing with a cigarette as a reward for their efforts. Sometimes these busy, powerful men with their feet up were doctors. However, this was not a doctor trying to reassure you that the cigarettes would not kill you; this was 'one of the busiest men in town . . . on call 24 hours a day'; 'a scientist, a diplomat, and a friendly sympathetic human being all in one', taking a hard-earned rest and smoking for pleasure. These ads had everything; they broke the associative connection between smoking and health/ill health/mortality/death highlighted in the previous ads involving doctors, by using the doctor merely as a reassuring role model and an exemplar of the class of busy, successful men who deserve a break. In addition, the doctor is described as a 'scientist' no less (how many times have you heard a run-of-the-mill GP described as a scientist?). A scientist who was presumably capable of evaluating the accumulating evidence and then making a conscious, reflective choice to smoke Camels ('More Doctors smoke Camels than any other cigarette'). Interestingly, the capital 'M' in 'More' and the capital 'D' in 'Doctors' are in red (thus 'MD'), to use a perceptual grouping cue, to pull the letters 'M' and 'D' together, and to make them stand out from their background In other words, we are talking here about medical doctors, real doctors rather than PhDs or (even worse) quacks, proper medical doctors. These were powerful and effective ads in terms of their effects, that is to say their sales figures.

However, Dichter did something else in his work on the psychology of smoking and the effective marketing cigarettes that has rarely been commented on. He laid down a marker for future attempts to defend smoking and the promotion of smoking by suggesting that any evidence for a statistical relationship between smoking and ill health could very well be an artefact of something else. He wrote,

> Efforts to reduce the amount of smoking signify a willingness to sacrifice pleasure in order to assuage . . . feelings of guilt. The mind has a powerful influence on the body and may produce symptoms of physical illness. Guilt

feelings may cause harmful physical effects not at all caused by the cigarettes used, which may be extremely mild. Such guilt feelings alone may be the real cause of the injurious consequences.

(Dichter 1960)

In other words, it is not smoking that gives you the cancer; it is the guilt that you have about smoking (and this guilt originally derives from your parents attempting to censure this form of behaviour in their offspring, even when they themselves smoked). Don't blame cigarettes for your ill health; blame your parents! However, bear in mind that he had also said that the role of the advertiser was 'not so much to sell the product as to give moral permission to have fun without guilt'. He was going to help us have fun without guilt and, therefore, reduce the physical harm of smoking. He must have known this was a load of baloney even then.

I do find it incredible that in the 1960s and '70s ordinary people were not more alarmed by smoking. How could many ordinary people not see the dangers it posed? Perhaps it was because they were confused by the scientific evidence; perhaps they had their doubts. Apparently, not every scientist agreed with the research findings on the link between smoking and lung cancer, or so it seemed. I remember hearing Hans Eysenck, perhaps Britain's best known post-war psychologist, in the mid 1990s, say (even then!) that the apparent statistical link between smoking and lung cancer was really just an artefact of personality. He argued that certain personalities were prone to lung cancer and certain personalities were prone to cancer and it was this underlying personality dimension that was the significant factor here. It was not smoking that gave you lung cancer; it was your genetically determined personality which gave you lung cancer. In other words, he was arguing that the scientific research was confounded; it was, in effect, fatally flawed. Moreover, Eysenck had been doing this work since the 1960s. So what chance did my mother or my Aunt Agnes, who left school at fourteen, have? Eysenck was publicly disputing the scientific evidence linking smoking and lung cancer, successfully turning the issue into a 'debate' between scientific experts. Smokers could cling on to this 'uncertainty' to rationalise their own behaviour.

In 2011, Pettigrew and Lee carried out an extensive review of recently released tobacco industry documents (released, it has to be said, because of litigation) that reveal a great deal about how the industry fought back against the growing scientific evidence on the relationship between smoking and lung cancer. The tobacco industry wanted to open up a great 'debate' about the effects of smoking on health, to suggest that the medical evidence was far from conclusive, to show that there were differences of opinion amongst experts on this topic. The tobacco industry created the Council for Tobacco Research (CTR) in 1953 to fund research that could be used in this fight. One recipient of this funding was the distinguished scientist Hans Selye, the so-called father of stress (Selye 1976), who had extremely impressive academic credentials with

1700 articles and thirty-nine books to his name (and apparently nominated for the Nobel Prize ten times). Pettigrew and Lee discovered that it was Selye who first contacted the tobacco industry (rather than vice versa) as far back as 1958, seeking funding for his research on stress. This first request was not successful. However, the following year a law firm representing the tobacco industry now involved in litigation, wrote to Selye offering him US $1000 to write a memorandum demonstrating that 'medicine has previously seen striking correlations suggested as representing cause and effect, only later to find that the significance, if any, of the correlation was otherwise' (Pettigrew and Lee 2011: 412). Selye agreed to do this for the money offered, but only on the understanding that any quote used would not be attributed to him; neither did he want to appear as a witness in any court case. This, as they say, was just the start of a very long and rewarding relationship. Given his academic credentials, he was of enormous value to them. He was after all an objective scientist, or at least that's how it would have seemed to the public if they did not know about his financial connection with the industry. Selye advised the tobacco industry that it should defend itself by focussing on the 'prophylactic and curative' aspect of smoking. Smoking was to be marketed as a way of adjusting to a stressful lifestyle. Selye was prepared to argue that it is stress that kills rather than smoking. Smoking, he suggested, can actually help you cope with this stress; it was in fact beneficial. In 1969, Selye, according to Pettigrew and Lee, 'testified before the Canadian House of Commons Health Committee arguing against antismoking legislation, opposing advertising restrictions, health warnings, and restrictions on tar and nicotine' (2011: 413). He was now being funded to the tune of US $100,000 a year back in the 1960s (about three quarters of a million dollars in today's money). He appeared on the Canadian Broadcasting Corporation arguing for the benefits of smoking for those under stress. Smoking, he argued, was a 'diversion' to avoid disease-causing stress. Oddly, he failed to mention this conflict of interest in the broadcast; he failed to point out that the tobacco industry was paying him as a spokesperson.

In the UK in the meantime, Hans Eysenck was playing a similar role in this 'debate' about the harmful effects of smoking. He was another very high-profile and influential academic (indeed it was one of Eysenck's books that got me interested in psychology in the first place) who publicly disputed the link between smoking and health. Like Selye he was also secretly receiving money from the tobacco industry, from the 1960s onwards. There are those, of course, who argue that this money from the tobacco industry was just research funding, and that scientists like Eysenck have to find funding for their research from wherever they can. They argue that such funding need not necessarily hinder the scientific objectivity of the researcher. If you believe this, then it is perhaps taking the time to reread Eysenck's first book on this topic, *Smoking, Health and Personality*, first published in 1965, which was a significant year for tobacco in the UK in many ways. The *Report on Smoking and Health* had been published by the Royal College of Physicians in England in 1962, and this had warned of the close connection between lung cancer and smoking. The findings of this report were widely publicised in the press. These were now clearly critical times for the tobacco industry. Eysenck aimed to show that the results of all of this medical research were 'by no means immune to challenge'. At one level, this is perfectly reasonable. It is the duty of scientists to challenge the establish orthodoxy, to present alternative hypotheses, to question and probe. But I cannot read this book and feel that I am just witnessing an inquiring scientific mind in action. There is just something about the tone and the nature of the arguments that he presents that makes me extremely uneasy, and not just with the benefit of hindsight. This book goes way beyond reporting a psychological theory that could have implications for the research on smoking and health, it does much more than that in, what seems to me, a cynical and calculated manner. It is not like any psychology book that I have ever read.

Eysenck's hypothesis was that there are certain types of personality who are prone to cancer and that there are certain types of personality inclined to smoke. He argued that it was personality that was the critical and confounded factor in cancer risk (Dichter, of course, had said that it was 'guilt' that was the confounded factor; Selye said it was 'stress'; clearly money bought a lot of possible unaccounted for factors). Eysenck proposed that certain types of personality, namely extraverts, are drawn to nicotine because it is a stimulant and therefore has introverting effects – in other words, extraverts smoke for good, 'genetically determined' reasons. Of course, extraverts differ in many other ways as well to introverts, for example, they have a

> preference for coffee and alcohol, for spicy foods, for premarital and extramarital intercourse, their impulsive and risk-taking behaviour – all these can easily be deduced from this general hypothesis. We may similarly deduce from it that extraverts would be more likely to seek for the stimulation afforded by cigarette smoking, and it is on this basis that the original hypothesis was formulated.
>
> (Eysenck 1966: 75)

So already, you have the idea that many smokers may differ in a number of significant ways from non-smokers and that any differences in health, like cancer risk, may not be attributable to the act of smoking per se.

Then, Eysenck quotes the conclusions of Doll, who he describes as 'one of the scientists most prominently associated with the promulgation of the theory that smoking causes cancer' as saying

> When the nature of the disease makes it impossible to carry out logically conclusive experiments there is always room for honest difference of opinion. In the case of smoking it is particularly hard to envisage how a conclusive experiment could be carried out and no such experiments have been made.
>
> (Eysenck 1966: 12)

Eysenck then adds, 'Doll goes on to quote a famous saying of Claude Bernard, to the effect that 'There are no false theories and true theories, but only fertile theories and sterile theories' (Eysenck 1966: 12).

Eysenck here does a number of things more or less at the same time. He suggests that in the case of lung cancer, conclusive experiments cannot really be carried out and therefore 'there is always room for honest difference of opinion.' In other words, what he is about to present in the rest of the book is just an 'honest difference of opinion' but, in addition to this, he also rejects the idea that theories can either be true or false, instead he tries to replace this with the distinction between 'fertile' or 'sterile' theories. It is not that some theories are right and some are wrong, he is saying, it is just that some are more 'fertile' than others are, and what he is going to outline is a particularly new and fertile theory. In other words, he is really undermining the readers' confidence in their ability to know whether a theory is right or wrong, by suggesting that this neat distinction does not actually apply to scientific theories. Theories about smoking and cancer are not right or wrong, they are something else altogether!

He then goes on to undermine the connection between lung cancer and cancer by pointing out that in the past; smoking was blamed for a whole series of other ailments as well. He writes:

> Among the ailments blamed upon smoking were lunacy, cerebral haemor-rhage, paralysis, delirium tremens, laryngitis, bronchitis, dyspnoea, tuber-culosis, dyspepsia, gastritis, intestinal rupture, heartburn, hepatic lesions, diarrhoea, flatulence, impotence, baldness, typhoid, skin diseases and many others. The children of smokers were supposed to suffer from hypochron-driasis, hysteria and insanity.
>
> (Eysenck 1966: 16)

He adds that these accusations were based 'on no scientific evidence of any kind'. What he is saying here is that smoking has been blamed for a whole series of things, which, with the benefit of time and hindsight, turn out to be quite ridiculous. He is implying that the same thing might be true of lung cancer and smoking. What specific effect might this have on the reader? Well, nobody wants to look ridiculous (at least nobody I have ever met), so perhaps it is better not to jump to any conclusions about the connections between smoking and lung cancer or you might be in the same boat as those that have concluded that there is also a relationship between smoking and flatulence.

Eysenck then reviews the epidemiological evidence of the relationship between the risk of contracting lung cancer as related to the average amount of tobacco smoked, and the evidence seems to suggest from research published in 1961 that there is a strong correlation between smoking and Group 1 lung tumours, which are 'made up of epidermoid carcinomas, and small cell ana-plastic carcinomas', but no relationship between smoking and Group 2 lung tumours. Group 2 tumours 'are made up of adenocarcinomas and bronchial or alveolar cell types.' Again he deliberately tries to undermine the readers' confidence by saying that 'lung cancer clearly is not just one undifferentiated entity; we are dealing with presumably at least two and possibly more quite different types of disease and each of these different types has different relations to smoking' (Eysenck 1966: 20).

But his next attack is even more extraordinary. He says that: 'the problem with correlating risk with smoking is that there are many different variables associated with the act of smoking which impact on the precise chemical composition of the smoke.' Thus,

> Some of these variables are the intensity of the suction applied, the length of the pull, the length of intervals between pulls, and the particular part of the cigarette which is being smoked, i.e. whether it is the first inch or the last inch of the cigarette. Thus, what we in fact inhale depends very much on the way in which we smoke a cigarette, and it is quite impossible to generalize in any sensible way without knowing more about these different variables we have enumerated.
>
> (Eysenck 1966: 20–21)

Therefore, what he is suggesting is that there are potentially important factors connected to the act of smoking itself, which researchers have not controlled, and that this problem underpins all the science relating smoking and lung cancer. Thus, he writes:

> We can ask people how many cigarettes they smoke, and we may even get a fairly truthful answer. However, we cannot ask them just how they smoke these cigarettes because they themselves would be unable to give us a reasonable answer, and we cannot find out by observing them either because the very act of observation would make them change their pattern of behaviour. Under these circumstances all statistics relating to smoking must be regarded with considerable caution.
>
> (Eysenck 1966: 23)

In other words, here he is saying that any statistical data, which shows a relationship between risk and lung cancer, is faulty because we cannot, with any certainty, isolate what the critical variable is. It is not necessarily the number of cigarettes you smoke, it is the way you smoke them, and this lets every heavy smoker, with their own signature style of smoking, off the hook.

Another chapter is called 'Giving Up Smoking?' Eysenck poses the question:

> Is it rational for him [the typical smoker] to give up smoking or continue to do so? If he continues smoking heavily then he runs the risk of having at the end of 75 a shorter life than a non-smoker by roughly 1.4 years. He might reasonably reply that there are so many hazards involved in life in any case that this relatively short period of longer life, problematical as it is, would certainly not compensate him for the lack of pleasure that would be

involved in giving up smoking. Is this a rational or an irrational attitude? . . . Certainly many people when questioned give some such answer in explanation of why they don't give up smoking; the immediate loss of pleasure and satisfaction is not compensated for by the problematical increase of a year or two in their life span at the age of 70 or above.

(Eysenck 1966: 111–112)

He is saying explicitly that it is not irrational to continue smoking even if smoking were found to increase one's mortality.

In the final chapter, 'Where There's Smoke There's Fire', he argues:

The evidence also suggests, however, that atmospheric pollution is probably an even more important factor, and that it would be unwise to concentrate all available research efforts and legislative measures on smoking. It is psychologically much easier to cause people to give up those habits which lead to atmospheric pollution than to give up smoking, and if our aim is the lessening of the terrible toll which lung cancer takes of life nowadays this avenue seems to be the more promising to take.

(Eysenck 1966: 116)

Put your efforts elsewhere, he is saying; leave those nice tobacco companies alone. However, we now know that Eysenck received more than £800,000 through a secret US tobacco fund called Special Account Number 4 (*The Independent*, 31st October 1996). Eysenck was, of course, not alone, in secretly taking this money, as we have already seen.

Oreskes and Conway, in their excellent 2010 book *Merchants of Doubt*, describe how on the 15th December 1953, the presidents of four of America's largest tobacco companies – American Tobacco, Benson and Hedges, Philip Morris and US Tobacco – had met with John Hill, the CEO of the public relations firm Hill and Knowlton, at the Plaza Hotel in New York. Their aim was to challenge the scientific evidence that smoking could kill you. In the words of the authors:

They would work together to convince the public that there was 'no sound scientific basis for the charges,' and that the recent reports [about cigarette tar and cancer] were simply 'sensational accusations' made by publicity-seeking scientists hoping to attract more funds for their research. They would not sit idly by while their product was vilified; instead, they would create a Tobacco Industry Committee for Public Information to supply a 'positive' and 'entirely pro-cigarette' message to counter the anti-cigarette one. As the U.S. Department of Justice would later put it, they decided, 'to deceive the American public about the health effects of smoking'. At first, the companies did not think they needed to fund new scientific research, thinking it would be sufficient to 'disseminate information on hand'. John Hill disagreed, 'emphatically warn[ing] . . . . that they should . . . sponsor additional research', and that this would be a long-term project. He also suggested including the word 'research' in the title of their new committee, because a pro-cigarette message would need science to back it up. At the end of the day, Hill concluded, 'scientific doubts must remain.'

(Oreskes and Conway 2010: 15–16)

Eysenck is not mentioned in the book on those who marketed doubt for a living. But his work, along with Selye and many others, contributed greatly to the uncertainty about smoking and cancer, and again in the words of Oreskes and Conway, 'throughout the 1950s and well into the 1960s, newspapers and magazines presented the smoking issue as a great debate rather than as a scientific problem in which evidence was rapidly accumulating.' My Aunt Agnes liked to read about the debate in the papers, and perhaps it was all hypothetical anyway for her. The statistics on those who worked in the carding room suggested that starting with the age of seventy-five and deducting a few years was always going to be unlikely in the first place. 'Did you know?' she said one day, 'scientists used to think that smoking gave you wind. I think that's bloody stupid. It's the opposite if anything. It's always the non-smokers who are the worst in the wind department.'

Then there was all that pleasure and satisfaction from smoking. 'The only pleasure we get around here,' in my mother's words. I can recall old cigarette ads from my childhood with a mixture of nostalgia and revulsion – these are fragments of memories of my childhood. I research some others from that time. I watch cigarettes square dancing, healthy, vigorous and perhaps most impor-tantly, connected. The cigarettes are the people in this American advert and they are all moving as one. 'Are you feeling alone, isolated?' the implicit message in the ad says 'have a cigarette and you will connect'.

I'm now sitting on a beach in Santa Barbara in California, and I watch one good-looking blond Scandinavian teenager, perhaps eighteen or nineteen, approach two Californian girls of approximately the same age, lying there on the beach in pink bikinis. He asks them for a light and he bends down as they pull a lighter out of their beach bag and light his cigarette, and they make a connection instantly and effortlessly. How else could he have achieved this? What else could he have said or asked for? 'Have you got a map? Do you have a compass? Do you have a water bottle? Do you have a syringe? Have you got some after sun? None of it would have worked. However, there was always the slick and easy 'have you got a light?'

I have on that beach in front of me a classic Marlboro ad from the 1970s (reproduced and analysed in Bullock 2004). The ad features a cowboy on a horse rounding up some stray horses. There are several outlines of horses; some can be clearly seen. However, when you look closely, Bullock argues, things are not what they might at first appear. The horse he is trying to round up is very feminine in a human sort of way, with distinctive female features, soft and yielding. However, the horses in the background of the advert are the most significant. I trace their outlines carefully and I realise that they are not horses at all but wolves, wolves that have been embedded in the scene. You need to look carefully to see them. Psychologists have studied these sorts of ambiguous visual stimuli for years, and this is one such visual illusion, there are really two figures in one. The idea is that these ambiguous figures are open to different interpretations but only one can be in consciousness at any one time. We normally see the figures as horses but really they have the outline of the wolf, and our unconscious mind can pick this up. Nevertheless, these hidden interpretations can impact on the brain. The wolves in the background remind you implicitly and unconsciously of loneliness (the lone wolf) and induce a degree of despair (after all, the wolves are bearing down on you), but not enough for you to consciously notice. How do you reconnect with other human beings, and how do you cope with the anxiety that is being induced at that point in time? That in a sense is the material essence of the ad, and a cigarette is the solution to both problems. A cigarette will allow you to connect with others who signify their identity through smoking, and a cigarette will allow you to use a form of infantile behaviour, like sticking something in your mouth, like a teat, or a thumb to calm your nerves, but all in a socially acceptable way to deal with the anxiety.

We all know that smoking is inherently unhealthy; they are coffin nails, Duck was right back then, even if he said it as a joke, so why did he and all the others smoke? Was it partly because they still had a little seed of doubt at the back of their minds about how dangerous cigarettes really are, a seed planted by the tobacco firms working in conjunction with credible scientists, who were on commission and somehow failed to declare this extraordinary conflict of interest to the public? There might be more to it than this. Ernest Dichter saw to that. Smoking, it seems, as much as anything else is driven by our great fear of being alone, a great fear deliberately induced and manipulated by insightful psychologists from the private psychology labs in the United States all those decades ago. Cigarettes were marketed to help us connect ('You're never alone with a Strand'), partly building upon that great unconscious symbolism of the connecting influence of fire and partly on the sharing of fire, pleasure and the relief of tension. Cigarettes may indeed help us connect, but not in a good way.

I think back to my Aunt Agnes, in the mill and their smoke breaks, away from the dust of the carding room and the machines where the factory girls could chat in relative quiet and reconnect and clean their lungs with menthol cigarettes when they could afford them. They could feel that they would never be alone with a Strand, or even a Park Drive or an Embassy, and they weren't. My Aunt Agnes was in her early sixties when she died, and many of the girls from the carding room died at that sort of age. But, at least, they were all together, united, the carding room girls, the big smokers, great friends, my mother always liked to say, and that is one way of looking at it. However, it is not my way.

The ad men who had some special insights brought the unconscious mind into focus, and then targeted and manipulated this unconscious mind relentlessly in pursuit of profit. It was a direct attack on the unconscious mind (whilst it was still out of favour in academic psychology), and like any conflict, one side eventually won, and, of course, it was not the side of the ordinary man and woman who emerged unscathed. The manufacture of doubt was also a critical component.

They fell for it back then, and they still do; and to me that is an awful tragedy.

- Ernest Dichter applied psychoanalytic methods to the marketing of cigarettes.
- He was not afraid to consider and target the unconscious mind.
- Dichter argued that you need to see below the surface, and *access* the unconscious both to cure neurosis and to *manipulate* the unconscious to sell brands.
- He suggested that if you want to understand how people think and feel about products, you cannot just ask them directly – psychoanalysts would never do that, it would be ludicrous ('please tell me about your neurosis'); you need more indirect and more intense methods.
- You need the psychoanalyst's couch, but on an industrial scale.
- You do not take what they say at face value; you need to understand instead the symbolic importance of products in people's lives, and you need to interpret what they say.
- You need to look out for defence mechanisms and projection.
- Dichter wrote that 'a number of human motivations are irrational, unconscious, unknown to the people themselves. This principle means that most human actions have deeper motivations than those which appear on the surface, motivations which can be uncovered if the right approach is used.'
- He argued that, from a psychological point of view, cigarettes work in a number of distinct ways – they allow you to behave like a child again, able to 'follow your whims'.
- They also offer a 'legitimate excuse for interrupting work and snatching a moment of pleasure'. Like children, we crave rewards – 'a cigarette is a reward that we can give ourselves as often as we wish.'
- Dichter argued that they should use this insight of self-reward as the basis for a marketing campaign.
- However, that was only part of his observations about the nature of smoking. Some of his respondents had also commented that with cigarettes, you never really feel alone.
- This was tapping into the primitive concept of fire, a warm glow, a conditioned stimulus rooted in our evolutionary past associated with the group assembling around the fire. One of Dichter's respondents said that 'when I see the glow in the dark, I am not alone any more,' and Dichter added that the use of cigarettes to combat feelings of loneliness and isolation was critical.

- Dichter wrote that 'the companionable character of cigarettes is also reflected in the fact that they help us make friends.'
- This insight formed the basis of a number of marketing campaigns.
- He also recognised the oral pleasure that derives from smoking, 'as fundamental as sexuality and hunger'.
- Smoking is about oral gratification, as infantile in the context of Freudian theory as sucking your thumb.
- Smoking is a socially acceptable way of obtaining this oral gratification by simultaneously sending out a powerful message about virile maturity and potency.
- Smoking is *so* powerful because it simultaneously satisfies infantile desires and yet symbolically sends out a very adult signal.
- Dichter argued that we have to deal with the essential psychological conflict that smoking generates.
- He wrote, 'One of the main jobs of the advertiser in this conflict between pleasure and guilt is not so much to sell the product as to give moral permission to have fun without guilt.'
- He also argued that it is not smoking that gives you the cancer; it is the guilt that some people have about smoking that causes cancer. This is clearly false.
- Dichter implied that if you can actually smoke guilt-free ('fun without guilt'), then you will be fine.
- This was part of a widespread and systematic effort by the tobacco firms to generate doubt about the relationship between smoking and cancer.
- On the 15th December 1953, the presidents of four of America's largest tobacco companies met with the CEO of a public relations firm to challenge the scientific evidence that smoking could kill you.
- Their goal was 'to convince the public that there was "no sound scientific basis for the charges", and that the recent reports [about cigarette tar and cancer] were simply "sensational accusations" made by publicity-seeking scientists'.
- The manufacture of doubt in this way was critical.
- Hans Eysenck, one of Britain's foremost psychologists, suggested that the link between smoking and cancer was really an artefact of personality – certain types of personality are prone to stress and smoking, and also prone to cancer.
- Eysenck received more than £800,000 for this research through a secret US tobacco fund called Special Account Number 4.
- Many people, including members of my family and my friends, bought into the glamour of smoking, and were reassured by those esteemed scientists, these 'merchants of doubt', in Conway and Oreskes' words.
- Many of them died through smoking-related diseases.

9 Our house is on fire

I was staring at a television screen. There was a young girl in a casual grey checked top holding some notes in front of her; her hair was in pigtails. She was seated. She looked younger than her sixteen years. The screen behind her read 'World Economic Forum'. It was the incongruity between the caption and this image of a child who hadn't bothered to dress up for the cameras that grabbed one's attention. There was silence, she adjusted the microphone and then she started to speak. She sounded confident, she didn't smile. 'Our house is on fire . . . I'm here to say our house is on fire . . .' She paused after each short sentence, the sentence was repeated, her eyes flicked to the left and right as if monitoring for feedback but you felt that any feedback wouldn't influence either the message or the delivery. It was a wary sort of watching. Then there was another pause.

> According to the IPCC we are less than twelve years away from not being able to undo our mistakes. In that time unprecedented changes in all aspects of society needs to have taken place, including a reduction in our CO_2 emissions of at least fifty percent.

I was still staring at the screen. There was no doubt in her message, no uncertainty, it was not couched in that probabilistic terms of scientific discourse. Science necessarily relies on hypotheses and probabilities, and it endlessly uses terms like 'highly likely' and 'extremely likely' when it comes to climate change. The Intergovernmental Panel on Climate Change (IPCC) themselves say things like

> it is highly likely that human beings have contributed to climate change through their behaviour, on the basis that changes in greenhouse gas emissions and global warming have mirrored major changes in human activity, like the industrial revolution, and changing patterns of land use, energy demands and transport.

The public don't like these terms; they don't really understand them. That is part of the problem of why they don't engage more with this issue. These terms suggest uncertainty, disagreement and doubt but in reality it is quite

DOI: 10.4324/9781003282051-9

the opposite – there is a *remarkable* scientific consensus on climate change – 'remarkable' because it is rare to see this degree of scientific agreement on anything. Science is, after all, fuelled by dispute, disagreement and difference. That indeed is its nature – that's how it develops and grows, and *changes*. But when it comes to climate change, the scientists *agree* that there's been an increase in greenhouse gas concentrations in the atmosphere and that this is linked to a general warming of the planet. There is agreement that mean temperatures have increased over the past century, and that they will continue to grow and then the crunch – 'highly likely that human beings have contributed to climate change through their behaviour'. But it's not 'certain', the critics say, not like death or taxes, as Benjamin Franklin wryly noted (the only things in life that are really certain); it sounds woolly and vague to people unused to probabilistic reasoning. Climate scientists also agree that the impacts of climate change on the planet will be severe, but with variability in exactly how severe. They have modelled a range of possible outcomes, but working out the exact probability of each possible outcome is more problematic because of degrees of uncertainty in the modelling, including knowledge of the earth's climate system and . . . future human activity.

That's the problem with science, it frames everything in probabilities and likelihoods and hints at possible doubt, but the girl with the pigtails and the odd sweeping eye gaze doesn't.

The IPCC, which comprises hundreds of the world's leading scientists, is the international body charged with reviewing and evaluating the vast body of accumulating scientific evidence around climate change. It was awarded the Nobel Peace Prize in 2007. Over the past three decades, it has issued a succession of reports and 'consensus statements' summarising the current state of extant knowledge on climate change, with the accumulating evidence, still couched in probabilistic terms, pointing more and more to one inescapable conclusion. In 1996, the IPCC concluded that 'the balance of evidence suggests a discernible human influence on the global climate.' In the 2007 report, the IPCC concluded that

> human activities . . . are modifying the concentration of atmospheric constituents . . . that absorb or scatter radiant energy. . . . Most of the observed warming over the last 50 years is very likely to have been due to the increase in greenhouse gas emissions.

In the 2013 report, the IPCC concluded that 'warming of the climate system is *unequivocal* [italics added] and since the 1950s, many of the observed changes are unprecedented over decades to millennia. . . . It is extremely likely that human influence has been the dominant cause.' In 2015, the IPCC concluded that they are 'now 95 percent certain that humans are the *main cause* of current global warming' (IPCC 2015: v; italics added). The IPCC also suggested that on the basis of the existing evidence that a rise in global temperature will have 'severe and widespread impacts on . . . substantial species extinctions, large risks to global and regional food security . . . growing food or working outdoors', as well as producing more extreme fluctuations in weather, including droughts, flooding and storms. The conclusions of the IPCC have been endorsed and supported by over 200 scientific agencies around the globe, including the principal scientific organisations in each of the G8 countries, like the National Academy of Science in the United States and the Royal Society in the UK.

Furthermore, an increasing number of people have been witnessing the devastating effects of climate change first-hand, with increased adverse weather conditions such as frequent flooding, stronger hurricanes, longer heatwaves, more tsunamis and periods of drought (IPCC 2015; UK Climate Change Risk Assessment 2016). The World Health Organization (WHO) (2017), warns that with temperatures rising and the increase in rainfall, we need to be prepared for more illnesses resulting from climate change, including mosquito-borne infections like malaria, dengue and the Zika virus. The WHO report that 'Climate change already claims tens of thousands of lives a year from diseases, heat and extreme weather', and they say it is 'the greatest threat to global health in the 21st century.' Indeed, the World Economic Forum identified climate change as *the* top global risk facing humanity, a greater risk than weapons of mass destruction and severe water shortages (Global Risk Report 2016).

The evidence suggests that human beings are the most significant contributor to climate change through energy use, population growth, land use and patterns of consumption (IPCC 2015). Currently, CO_2 emissions from human activity are at their highest-ever level and continue to rise. Global CO_2 emissions in 2011 were reported as being '150 times higher than they were in 1850' (World Resource Institute 2014; see also IPCC 2015). Although we cannot undo the damage already done with regards to climate change, we do have the power to adapt our behaviour to ameliorate any future effects (Beattie and McGuire 2018; see also Beattie 2010).

Although the role of human activity in its causation is both 'clear' (and 'growing'), evidence for large-scale behavioural adaptation on the part of the public is absent. Indeed, there appears to have been a monumental disconnect between the science of climate change and the public's perception of climate change and their subsequent actions over the past decade or so. For example, a 2013 survey by Yale University had found that only 63% of Americans 'believe that global warming is happening'. Interestingly, this figure had been higher (72%) back in 2008 before the effects of the economic crisis were fully felt, and before the 2009 'Climategate' scandal where emails of climate scientists at the University of East Anglia were hacked. It was suggested at that time that there was some manipulation of the scientific data and that climate scientists, like everyone else, in this great 'climate change debate' had a vested interest to protect. Belief in climate change dropped to 52% in 2010. Nearly half of Americans in a 2010 survey thought that global warming was attributable to natural causes rather than to human activity – climate scientists clearly think otherwise. The girl in the pigtails aimed to put things right.

But it was now 2019 and I was watching the news bulletin in my office at the university with two others: one was a young student, here for a tutorial; the other was a very successful marketer. The second was at the university to talk about how insights from psychology could be applied to the real world of business. I had told them that I wanted to catch this particular bulletin. We turned the bulletin on just in time. The young girl on the screen had started to speak. The nonverbal response from the two in the room to what they were watching could not have been more different; I could see it in their faces: one was smiling and nodding her head, the other was frowning and making a blowing sound when the young girls finished. She screwed up her face and then glanced at me for support.

Sometimes you don't have to ask people what they think about something, but I did anyway.

The student jumped in first. 'I'm her greatest fan,' she said. 'It's about time politicians had a rocket up their . . .' She didn't finish the sentence in front of our distinguished visitor. Our distinguished visitor frowned again, and then she started to talk. 'What are we supposed to do?' she asked.

> Close all businesses? Shut down the world? I'm not sure that telling us 'Our house is on fire' is actually very helpful. In my work we're greening so many parts of the economy, we're working flat out on this, I'm working with multinationals on this and with governments, but she's basically saying that this won't do. She's not encouraging me to try harder, she's not encouraging me to believe there's anything I can do. She's just trying to frighten us and make me doubt my own competence to do anything.

'But she's good at standing up to the politicians,' said the student enthusiastically, 'and telling them what to do. It's the right image. It's terrifying. That's exactly how I feel. Our house is on fire – that's why I get so angry about it.'

'You mean telling them to run out of the burning house,' said the marketer, 'and abandon the lot of us. You can see why some people try to bury their heads in the sand or try to cling on to opposing views.'

The marketer had got angry, which she tried to soften with a light laugh at the end. But we had heard her frustration and the student's anger.

Doubt has always been central to the climate change issue; it can certainly be the harbinger of inaction with respect to taking steps to mitigate the effects of climate change. If there's some doubt about whether it's real or not, and some doubt about the role of human beings in causing it, then these are great personal justifications for *inaction*. Why would you buy an electric car or use public transport if you think that there is serious doubt about whether the whole thing is real? And the scientific discourse of climate change with its emphasis on likelihood doesn't help. It suggests lack of certainty, and that is enough for some people who label it 'fake news' – overblown, overstated, ideologically driven. There have been many climate change 'sceptics' over the past few years, not least Donald Trump and his constant 'fake news' message, which certainly played very well both in his election campaign of 2016 in those states which had been decimated by the decline of the coal industry and also after he became president. It was a message that the ex-miners and others wanted to hear. It reassured them about this so-called existential threat, it reassured them about their livelihood and it helped them feel good about themselves – all at the same time. No longer sheep, no longer following the herd. The message got global coverage. Trump had been tweeting about climate change for years. On the 1st November 2012 he tweeted, 'Let's continue to destroy the competitiveness of our factories & manufacturing so we can fight mythical global warming. China is so happy!' On the 15th February 2015 he tweeted, 'Record low temperatures and massive amounts of snow. Where the hell is GLOBAL WARMING?'

Closer to my home in Northern Ireland, the Democratic Unionist Party Minister Sammy Wilson was saying that he thought that 'man-made climate change was a con.' In the *Belfast Telegraph* (31st December 2008), he went on to say that 'there is now a degree of hysteria about it, fairly uninformed hysteria I've got to say.' He suggested that those who argued for climate change didn't understand the science and that they couldn't explain the 'connection between CO_2 emissions and the effects which they claim there's going to be'.

It was as if there were two tenable positions, two sides, both supported by the science. You can therefore choose which side to take – that was the implicit message. The BBC and other broadcasters had a 'duty' maintained over many years to have both perspectives represented in any 'balanced' discussion on climate change. This media balance reinforced the notion that there was serious doubt about the science of climate change, and many people liked this message (no existential threat, no change necessary, continue as before) and climate change sceptics seemed to be everywhere. Greta Thunberg was trying to remove all doubt with her simple message, with no ambiguity and no window dressing. But the problem is that too much fear in any message ('our house is on fire') is also not an effective way of gaining compliance – if the recipient of the message doubts (that word again) whether there is anything they can do to resolve the threat and reduce the fear (as many people do). You need to know that you have the power to do something (self-efficacy) and that your response will make a difference to mitigate the effects (response efficacy). If you have doubt about either or both of these things, you find other ways of dealing with the fear. Greta Thunberg now crystallised (and personified) this extreme fear message about climate change, but ordinary citizens often felt powerless about whether they could do anything which would actually make a difference. They developed an aversion to Greta and her message. In situations like this they may well attempt to avoid the message altogether – refusing to read articles about climate change or watch news items about it (or even turn off the TV when she appears on screen), or more subtly not attending to certain parts of the article even when they're right in front of you on a computer screen – these 'certain' parts to be avoided being the remarkable consensus on the science and the scary consequences. If the climate change sceptic arguments are presented as well, that is, those arguments that raise severe doubts about climate change, then you may find solace in these (Beattie 2018b). You may attend to these sections – your eyes automatically and seemingly unconsciously drawn to them. That is the information you then process to build your representation of the situation, to confirm your long-held views – namely that climate change won't affect *you* – just other people, overseas or in the future (or preferably both). You may even conclude that these climate change are an attempt by *others* to manipulate you (Beattie and McGuire 2018). These 'others' include Greta Thunberg herself, the IPCC, and various governments around the world. Those proselytising about climate change have been labelled by some (including famously Donald Trump) to be master manipulators, part of a great Chinese or Communist conspiracy to harm the West and Western economies.

In our own research on this topic, we demonstrated one type of anticipatory avoidance in an experimental psychology study where we asked university staff and students to read a series of climate change articles – articles about climate change in general, about climate change and its relation to flooding in the UK and about climate change and its consequences for food scarcity and violent conflict (Beattie 2018b; Beattie et al. 2017; Beattie and McGuire 2018). Each climate change article contained three arguments for climate change ('for') and three arguments against climate change ('against'). The 'for' arguments reflected the scientific consensus that climate change is real, that human activity is the cause of both climate change generally and flooding in the UK, as well as the predictions that climate change will cause food scarcity and conflict. The 'against' arguments were essentially the climate change sceptic arguments: that climate change is not occurring or is exaggerated and that it's not caused by human activity, that flooding in the UK is not caused by climate change, and that there is no link between climate and food scarcity and conflict. In other words, that there is serious *doubt* about the science. All arguments were drawn from print and electronic media (e.g. the *Guardian*, the *BBC News* website) and online blogs. The 'for' arguments came from news articles summarising the findings detailed in various IPCC reports about climate change. The 'for' and 'against' arguments were carefully edited such that they were of similar word count and frequency.

We used eye tracking to plot the individual eye-gaze fixations as our participants read the messages. We measured fixation count (number of individual fixations), fixation duration (the overall and mean duration of these fixations in milliseconds) and dwell times (overall time spent focussed on the different sections, measured in seconds) to 'for' arguments and to 'against' arguments, for both optimists and non-optimists.

We found that a significant proportion of our highly educated participants skipped over the damning scientific evidence about climate change, and the implications of climate change for our planet, and focussed instead on any threads about uncertainty in the science (the articles were, of course, designed to be balanced in this way). In fact, they spent half of their total time focussing on these sections – the sections highlighting *doubt* about climate change.

Those with a sunny, optimistic personality (measured on a particular psychometric scale called the LOT-R to assess 'dispositional optimism'; Scheier and Carver 1985) were significantly more likely to behave in this way – to focus on the (emotionally!) 'positive' sections of the articles – those sections suggesting that there is still some significant doubt about climate change (Beattie et al. 2017). Fixation durations were significantly shorter for optimists to the sections outlining the science and its implications than for the 'doubt' sections. These dispositional optimists like to maintain their positive mood state by avoiding negative emotional information, images or arguments that might dampen their mood state (Isaacowitz 2005, 2006; Beattie and McGuire 2011). In our study published in 2017, we found that *non-optimists* had longer fixations on the emotionally negative (but scientifically correct sections) about climate change and its implications for the planet.

The individual scan paths of two participants (one optimist and one non-optimist) are displayed below as they read arguments both 'against' and 'for' climate change (Figure 9.1). In this scan path, circles represent individual fixations on words, with larger circles representing longer fixation durations. Lines between circles represent saccadic eye movement behaviour. The text of the 'for' and 'against' arguments were grouped into Areas of Interest.

Figure 9.2 displays hotspot analysis of eye-gaze fixations of the group of optimists and non-optimists reading arguments 'against' or 'for' climate change. In this figure, greater intensity represents longer dwell times at fixated locations.

When participants were asked to summarise the articles after they had read them, the non-optimists were more likely to frame their subsequent account of the articles in terms of the evidence *for* climate change ('this article is about global warming and how 95% of it is due to human activity'). Two-thirds of their recalls were framed in this way. The *dispositional optimists*, on the other hand, who fixated significantly less on arguments for climate change but on sections about the doubt and 'uncertainty' were more likely to remember the articles in terms of a debate between two opposing positions ('it's about climate change, about trying to understand what's happening with the weather and there are different points of view'). Two-thirds of their recalls were framed as a debate. In other words, what we remember about climate change from articles on the subject is affected by our personality which influences our moment-to-moment fixations on positive or negative aspects of the messages. Dispositional optimists are drawn to doubt in this instance.

We also considered the relationship between level of dispositional optimism and the extent of so-called optimism bias – this is a bias where you overestimate the probability of good things happening to you in life (and underestimate the probability of bad things happening). According to Tali Sharot (2012) around 80% of people suffer from some form of optimism bias in many aspects of their lives – apparently believing that their marriages will work (it's only *other* marriages that fail, they say), their start-up businesses will succeed, and that they will have a long and fulfilling life compared to everyone else. This sort of unrealistic optimism would seem to be somewhat pervasive, affecting not just our personal relationships but also our attitudes to finance, work and health. For example, adolescent smokers are two and a half times more likely than non-smokers to doubt that they personally would ever die from smoking even if they

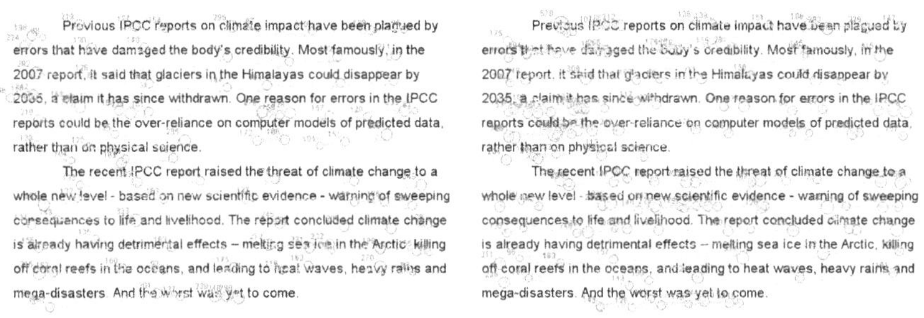

Previous IPCC reports on climate impact have been plagued by errors that have damaged the body's credibility. Most famously, in the 2007 report, it said that glaciers in the Himalayas could disappear by 2035, a claim it has since withdrawn. One reason for errors in the IPCC reports could be the over-reliance on computer models of predicted data, rather than on physical science.

The recent IPCC report raised the threat of climate change to a whole new level - based on new scientific evidence - warning of sweeping consequences to life and livelihood. The report concluded climate change is already having detrimental effects – melting sea ice in the Arctic, killing off coral reefs in the oceans, and leading to heat waves, heavy rains and mega-disasters. And the worst was yet to come.

Figure 9.1 Individual scan paths of (a) an optimist and (b) a non-optimist, as they read one argument 'against' climate change (first paragraph) and one argument 'for' climate change (second paragraph).

Source: Copyright 2019 from *The Psychology of Climate Change* by Geoffrey Beattie and Laura McGuire. Reproduced by permission of Taylor & Francis Group, LLC, a division of Informa plc.

Figure 9.2 A hotspot analysis of eye gaze fixations of a group of (a) optimists and (b) non-optimists reading one argument against climate change (first paragraph) and one argument for climate change (second paragraph).

Source: Copyright 2019 from *The Psychology of Climate Change* by Geoffrey Beattie and Laura McGuire. Reproduced by permission of Taylor & Francis Group, LLC, a division of Informa plc.

smoked for thirty of forty years; adult smokers are three times more likely to believe this. When it comes to smoking or climate change this optimism bias can have deadly consequences. Optimism bias has been found across a range of environmental issues (Gifford et al. 2009), as well as in estimates of the risk of health damage from specific environmental hazards, like water pollution (Pahl et al. 2005), and with climate change (Gifford 2011). A large, eighteen-nation survey demonstrated that individuals believe that across a number of environmental issues they are safer than others living elsewhere and that they are safer than future generations – in other words, they show both a spatial and a temporal bias.

Optimism bias appears to be associated with specific cognitive biases in processing relevant information. One study in behavioural neuroscience used functional magnetic resonance imaging (fMRI) to measure brain activity as participants estimated their probability of experiencing a range of negative life events, including things like Alzheimer's and burglary (Sharot 2012). After each individual trial, participants were then presented with the average probability of that event occurring to someone like themselves. The researchers found that their participants were significantly more likely to change their estimate *only* if the new information was better than they had originally anticipated. This bias was reflected in their fMRI data in that optimism was related to a reduced level of neural coding of more negative than anticipated information about the future in the critical region of the frontal cortex (right inferior prefrontal gyrus). They also found that those participants highest in dispositional optimism were significantly worse at tracking this new *negative* information in this region, compared to those who were lower in dispositional optimism. In other words, the optimism bias derives partly from a failure to learn systematically from new undesirable information, and this bias was most pronounced with those highest in dispositional optimism.

In our study, we found that optimism bias is significantly affected by underlying level of dispositional optimism as measured on that simple scale – for example, *optimists* in our study reckoned that they had a 36.5% probability of being *personally* affected by climate change, whereas they thought that *other people* had a 52.8% probability of being affected and that 76.4% of future generations would be affected. For *non-optimists*, the figures were higher throughout – 56.8% thought that they would be personally affected, 68.5% thought that other people would be affected and 84.1% thought that future generations would be affected. Even non-optimists, it seems, have some degree of optimism bias – they thought that they would be less likely to be personally affected by climate change than other people elsewhere and in the future. But the optimists in this sample were particularly blasé (about a one in three chance, they reckoned, that they would be personally affected – less than 50:50, a figure which may have major symbolic psychological significance).

These empirical results are perhaps more worrying when you remember that successful entrepreneurs and business leaders tend to be (highly resilient) optimists – in fact some have argued that it is the *key* psychological factor in achieving entrepreneurial success (Crane and Crane 2007). It may also be worrying given that there is a major self-help industry devoted to training people to be *even more optimistic* (Beattie 2011). For the past thirty or forty years, we have been striving to increase optimism in society because of its health benefits (through both positive psychology and a cultural emphasis on 'the power of positive thinking'). Optimism is, after all, highly advantageous for the individual, as Martin Seligman has consistently argued, because it has significant effects on both mental and physical health (Seligman 2002) and was selected for during evolution (Mosing et al. 2009). There is evidence that optimists live significantly longer and are much less likely to die from cardiac arrest (Scheier et al. 1989); it also increases the survival time after a diagnosis of cancer (Schulz 1996). Optimism does this by reducing stress and anxiety about the future, and optimists consequently have better immune functioning (Segerstrom et al. 1998). Belief in a positive future also encourages individuals (in *some* domains, particularly those that they have some control over) to behave in ways that can actually contribute to this positive future, thus becoming a self-fulfilling prophecy (Sharot 2012).

Although underestimating future negative life events can reduce our stress level and add to our longevity, sometimes actual negative events (like climate change) really do have to be considered. Optimism bias is potentially very dangerous when it comes to the discounting of serious risk. Some critics of 'the power of positive thinking' have argued that we have produced a profound and dangerous socio-psychological change in Western societies with unrealistic expectations about the future (Ehrenreich 2010). They have also argued that it has actually 'undermined preparedness' to deal with real threats like global terrorism, financial bubbles or climate change, with the public having 'no ability or inclination to imagine the worst'.

Optimism can be a positive thing (especially at the level of the individual), but like everything it has its limits. Over-optimism can be very damaging, and perhaps it is time to reevaluate this overarching cultural focus and consider new ways to get the public to consider or imagine the worst – whenever and wherever appropriate.

Of course, it is sometimes pretty obvious when people try to bury their head in the sand. But sometimes it's a bit less obvious, and our eye-tracking research suggested a different type of avoidance activity. It is not so much a head buried in the sand as the human brain actively searching for something – the positive emotional sections of these climate change articles. In the case of climate change, these positive emotional sections all relate to doubt about climate change, and scepticism and arguments about the exaggeration of the findings – it is a more or less complete counter-narrative. Donald Trump and other climate sceptics go one better – they make it a complete counter-narrative. They dispute not just the science but say that they can expose the architects behind this shady science, this 'fake news' – the IPCC ('mere puppets'), certain governments, the Chinese – all out to ruin Western business (and particularly out to destroy American business). We might not like Donald Trump or his politics (or political methods), but it is disquieting to think that at times we might read carefully the arguments his fellow travellers (if not Donald Trump himself) use about climate change to assuage our own emotional state, and perhaps we read them more carefully than the articles about the science and the consequences of climate change. Doubt, it seems, can be a great refuge.

But there is another factor that we have to consider. There seem to be deeply held cultural, political and religious beliefs that divide those who believe in the science of climate change and those who do not, and it is not *just* a question of science and scientific knowledge. This point was made by Stephen Pinker in conversation with Bill Gates a number of years ago, when he said:

> One of the biggest enemies of reason is tribalism. When people subscribe to an ideology, they suck up evidence that supports their preconceptions and filter out evidence that goes against them. Contrary to the belief of most scientists that denial of climate change is an effect of scientific illiteracy, it is not at all correlated with scientific literacy. People who believe in man-made climate change don't know any more about climate or science than those who deny it. It's almost perfectly correlated with left-wing versus right-wing orientation. And a move towards greater rationality would unbundle them and let evidence inform what the optimal policies ought to be.
>
> (Kasinger 2018)

The statistics on this divide fuelled by ideological position are striking. Andrew Hoffman, in his book *How Culture Shapes the Climate Change Debate*, reports that in 1997, 47% of Republicans and 46% of Democrats thought that climate change was already happening – in other words, virtually identical percentages. By 2008, the figures had diverged dramatically, with fewer Republicans holding this view (down to 41%) but with far more Democrats than previously expressing this position (up to 76%). By 2013, the respective figures were further apart still, 50% and 88%, respectively. Hoffman says that the cause of this polarisation on ideological grounds after 1997 was the Kyoto Protocol, which was the first international agreement to reduce greenhouse gas emissions, which was supported by the Clinton administration. Media attention on the political and economic implications of climate change rose dramatically in the years following. McCright and Dunlap (2011) reported that there were 166 documents critical of the science of climate change in 1997 alone. There were 107 climate change denial books published between 1989 and 2010. Most of these, according to Hoffman, were linked to conservative think tanks, and somewhat tellingly 90% did not go through a peer review process. They all raised the issue of doubt.

Hoffman's book acts a reminder of how economic (and political) factors and psychology are intimately connected, and that psychology, of course, is part of the world and not separate from it. The Kyoto Protocol to reduce greenhouse gas emissions had major implications for the energy sector and industry in the United States, and a counter-campaign was mounted. This brings us into murkier waters when we consider how both the science linking smoking and cancer and linking human activity and climate change were both turned into 'scientific debates', which allowed both viewpoints to flourish and stay literally worlds apart. The BBC and other media reinforced the concept of a 'debate' about climate change by having representatives from both sides of the argument. This might sometimes seem like the fairest position to adopt (notwithstanding the fact that the 'science' of climate scepticism had *not* gone through the peer review process, the very basis of science itself). One might argue that it allows the viewer to make up their own mind in a rational way when they are allowed to hear both sides of the argument. But this ignores the basic fact of the role of emotion (and mood) in directing human behaviour, and the importance of optimism bias and those desperate human attempts to stay cheerful in the light of incoming bad news. They were being confronted with an extremely fear-inducing message (indeed an existential message) without feeling that they were in a position to do anything about it. Balance in the debate here meant that doubt was paraded as fact, and instinctively and unconsciously many minds gravitated to this information with implications for what they remembered and how vulnerable they felt.

Doubt clearly can be weaponised. It can be very handy when you need to explain why you are personally not doing anything when your house is on fire, when you can say that you feel that there's nothing that you can do that will make any difference. We can cling on to doubt when we need to, because in certain situations it can be very reassuring. It can allow us to rationalise our instinctive emotionally directed behaviour as we direct our visual attention away from upsetting information (and what's more upsetting than the prognosis for our earth with further temperature rises?). It turns a clear form of emotional responding into something that looks like reasoned action and makes us appear as if we are in control of our behaviour, and we seem to like that feeling of apparent control.

Doubt is central to the climate change emergency and in many different guises. A scientific language couched in probabilities which hints at doubt where there is none; personal doubts about what we, as individuals can do and thus low self-efficacy as a result; counter-narratives driven mainly by large energy companies to allow us to view opposing 'scientific' positions, and anticipatory avoidance of the upsetting bad news about climate change by ordinary citizens as a result, even when that upsetting science is on a computer screen right in front of them. We gravitate to the doubts expressed in these articles, without any apparent conscious reflection, no matter how scientifically unsound they are. When it comes to climate change, doubt has clearly been weaponised, but this loaded weapon needs to disarmed. The science of climate change needs to be more unambiguously expressed with a clearer message about what *we* all can do – Greta's message was clear and unambiguous, but it didn't have the second part contained with it. We need to explain and show that we can all make a difference to greenhouse gas emissions (instead of merely leaving it up to others like the harangued politicians) and that the science of climate change has to be a stimulus for action for powerful educated people like us, to be considered carefully and responded to, rather than necessarily the harbinger of doom which must be avoided at all costs. If we doubt ourselves, we will always be drawn to any doubts about the existence of climate change.

It therefore seems that learning to be more effective at dealing with doubt may be critical in getting the public to face up to climate change, and hence critical for our very survival.

- Climate change is the biggest problem that we as a species have ever faced.
- Doubt is central to the climate change emergency and it appears in many different guises.
- There is a scientific language about climate change couched in probabilities which hints at doubt where there is actually none.
- Personal doubts about what we, as individuals, can do to mitigate the effects of climate change lowers self-efficacy and leaves us feeling powerless.
- There are counter-narratives on climate change, driven largely by large energy companies, presenting opposing 'scientific' positions on the subject and giving rise to doubt about its existence and its nature.
- There is often anticipatory avoidance of the upsetting bad news about climate change by ordinary citizens, even when that upsetting science is on a computer screen right in front of them.
- It seems that we gravitate to the doubts expressed in these articles, without any apparent conscious reflection, no matter how scientifically unsound they are.
- When it comes to climate change, doubt has clearly been weaponised, but this loaded weapon needs to disarmed.
- The science of climate change needs to be more unambiguously expressed with a clearer message about what *we* all can do.
- Greta Thunberg's message was clear and unambiguous, but it didn't have recommended actions for us all. It left severe doubt about our own self-efficacy for mitigating the effects of climate change.
- We need to explain and show that we can all make a difference to greenhouse gas emissions.
- If we doubt ourselves, we will always be drawn to any doubts about the existence of climate change to remain over-optimistic.
- If we do not defeat doubt here, our very survival is under threat.

10 Concluding remarks

This book has been an exploration of the concept of doubt, that most human of attributes. I recognised from the beginning that doubt, 'a lack of confidence or uncertainty about something or someone, including the self', is central to science, the law, ethics, politics and philosophy, but that it is also central to who we are. I stressed that it can be a safeguarding mechanism or a terrible distraction with a number of binary opposites pulling it this way and that – it can be rational or irrational, systematic or random, healthy or pathological. I was reminded that obsessive-compulsive disorder is often thought of as a 'disease of doubt'. I have attempted to explore some of these issues and some of the manifestations of doubt in people's lives. I have tried to take an idiographic rather than nomothetic approach to the subject and therefore I have used memoirs, autobiographies, biographies, personal letters, ethnographic observations of groups, sometimes combined with a bit of reflection and self-analysis. I have used the broader nomothetic psychological literature to help me understand and interpret what I have found.

Doubt is highly personal, and so we need an idiographic approach to get close enough to see it in action, to document it. Sometimes it is fully articulated by individuals, but sometimes it is not, and then it needs to be inferred on the basis of particular patterns of human action. I do this with the supremely confident Pablo Picasso. Good biographies can sometimes point the way towards the phenomenon and then we need to begin the process of analysis. Doubt has two sides, rational and irrational, which I have tried to show in some of the case studies that I've used, including Kafka, Jung, Picasso, Turing and Pauline Clance where sometimes we can see how the two sides of doubt, as a tool of rational thought and as an irrational distraction, are often intimately connected. One side can and does influence the other. Productive doubt can drive progress, but then more generalised doubt can inhibit it; indeed, it can then inhibit life itself. That is the beauty and complexity of doubt. But how could it be otherwise? One moment's reflection would tell you that those personal fleeting doubts that arise and zig-zag obstinately through our minds have different characteristics and sometimes can change form, and change in more substantive ways as well, as we reflect on them. I wanted great scientists and artists, including great scientists of the mind, to appear in the book because doubt sometimes drove

DOI: 10.4324/9781003282051-10

them on. But then that process of doubting spread and led to unforeseen and unforeseeable consequences – the rejection of Christianity (in the case of the psychoanalyst Jung), the rejection of self-worth (in the case of the psychologist Pauline Clance), the rejection of risk (Turing), the rejection of agency (Kafka), the rejection of science and a desperate clinging to magical thinking as an alternative form of prediction and control (in the case of Picasso). I wanted the book to document this connectedness of doubt but also how it can be addressed 'therapeutically', like in that gym in Sheffield – to give those young fighters a chance in the hard, cruel world of boxing. I also wanted to show how it can be manipulated by those with a vested interest in making us doubt when we shouldn't, for example, about the connection between smoking and lung cancer, or about the validity of the science detailing climate change. Both of these cases of the manipulation of doubt are, I think, particularly shameful episodes.

This book always had to be called a 'psychological exploration' because this was new and difficult territory. Doubt is highly personal, it's also a very broad concept and it's woven into people's lives. It looks like few people are entirely immune to doubt, and when you look carefully enough you can detect it in the lives of the most influential writers, artists and scientists. Jung viewed doubt as an indispensable component of a *complete* life – one of the most important drivers for human thought and action. Freud, on the other hand, viewed doubt as a symptom of obsessional neurosis deriving from the push/pull of conflicting instinctual impulses. Given the breadth of the concept, you can appreciate why such radically different views on the subject have arisen. I explored the role of doubt in Jung's own life and how he came to his conclusions. The origin of his own doubts was in a dream, and dreams – especially fantastical dreams – belong to the realm of the unconscious, which Jung explored with us in his autobiography. However, I did start this book's particular psychological (and geographical) journey in much more mundane and familiar settings – in a family home in Prague (although not such an ordinary household, with Kafka's 'repulsive tyrant' of a father), or even more mundanely up a hill outside Belfast in the driving spring rain trying to decide which subject to study in the fourth year of secondary school. My life is nothing if not mundane! But sometimes very mundane settings and lives give rise to all kinds of significant psychological issues (as we all know), and that hill is, I am sure, where my own personal problems with doubt first arose. Unlike Kafka, I didn't have anyone to blame my doubting on, there was no 'repulsive tyrant' in my home, and no oddity in my psychosexual development (cf. Freud), as far as I can tell. Just a very important decision in my life because of my family circumstances (poor, working class) and the fact that I had lost faith in experts who could advise me, because of my father's premature death on the blood-spattered plastic sheeting on a table in an operating theatre. That's when I first turned a decision, perhaps more important for me than my peers, into a conscious battle with doubt. Self-reflection tells me that I had stopped believing 'experts' because of his death – doctors, surgeons, perhaps my teachers, almost anyone in authority. Who then could advise me? I had begun to doubt: I found it harder to take advice. This seems to have stayed with me – and sometimes, this can be quite useful. I don't necessarily accept the opinions of important 'authority figures' at face value. I make up my own mind.

But unfortunately, my doubt did not stay in this one domain; it generalised to even trivial decisions – like the embarrassing procrastination over choosing a tin of peas in a supermarket, and a variety of what now look to me like obsessive-compulsive behaviours which started after the death of my father. I would take my scooter to the top gate of our park and ride to the bottom a set number of times. Always twelve times, counted out loud. Neither the weather, nor gangs on the street, nor the Troubles themselves could interfere with my nightly routine. Later, it was running, again compulsive, necessary, and again without any break. I have been stoned whilst out on my runs in Belfast, pelted with bricks, as I crossed one of those imaginary but nevertheless very real 'Peace Lines' that mark out working-class Belfast. No doubt I had some generalised anxiety underpinning my doubting brought on by that tragic loss in my early life, but it's that time on the hilltop that I now remember. The anxiety was instantiated and localised, fixed in time and place. It became part of our family's narrative, part of our folklore: 'Do you remember that time our Geoffrey couldn't make up his mind about biology or Latin?', and everybody would laugh. I had come back soaked through and I still wasn't sure what to study. It thereby became part of me. I went on to do a PhD and later became an academic, and like many academics, I felt like an impostor on many occasions. I keep it to myself but work harder to compensate. Despite what some people seem to think, impostor syndrome is found in both genders. The effects of impostor syndrome can be very debilitating – some, it seems, become workaholics, whereas some self-handicap to make sure that they never succeed. Its damaging psychological effects cannot be overestimated.

However, not everybody is riddled with doubt – Jean and Tom from Sheffield had several close encounters with aliens but no doubts whatsoever about what they had experienced. I looked to Festinger and his work on cognitive dissonance to help me understand this, and Wooffitt's work on the mundane reporting of odd or supernatural experiences where people act to construct the accounts of these experiences as both 'normal' and credible. Picasso didn't seem to have any doubts in his life, including any doubt about his greatness, but here I had to become something of a psychological detective and probe a little deeper into his behaviour and one or two of his (many) outbursts. I saw evidence of magical thinking here, a form of superstitious behaviour to keep doubt at bay. A separate belief system to keep him reassured and in control. Perhaps that's why his consciously expressed doubts seem to have disappeared, but his actions spoke volumes. Doubt was there but slightly obscured by his self-presentational strategies.

Alan Turing's genius is well documented and highly celebrated, but if his life is characterised by anything (as well as by terrible tragedy), it is psychological change and adaptation – a growth in self-efficacy occasioned by his academic and professional success as a great mathematician – this feeling that he would now succeed in whatever domain (and with probabilities attached by him to things going wrong). Turing eventually found himself in an unfamiliar social world, the world of gay cruising at the darker end of Oxford Road in Manchester with working-class lads trying to make a few bob, and this is when he perhaps needed some doubts and a better social understanding of people's motives and perceptions. We all know the tragic conclusion. But it is perhaps a story about where the removal of all doubt can lead you.

I wanted to know if doubt can be eradicated. I didn't go to what some might see as the experts – perhaps counselling or clinical psychologists; I went to see a sports trainer in action – Brendan Ingle, 'Professor of Kidology' (in his own words) from a poor background in Dublin, uneducated, but quite brilliant. He would make his young protégés line up and sing nursery rhymes in front of the group in the musty gym and face the banter and ridicule of their friends. Over and over and over again. This boxing culture was a lesson in change. I watched doubts disappear. Not necessarily immediately but over the years. This was a long-term observational study on my part (Beattie 1996). Now that they were free of doubt, a number became famous boxers, and one or two became world champions. But sometimes at a terrible personal cost – the level of arrogance in some of these fighters was frightening (Beattie 2002).

Then I considered how doubt can be weaponised first by the tobacco companies and then by the energy companies – generating doubt about the relationship between smoking and lung cancer and then doubt about the science of climate change. These were both murky tales with very significant societal consequences. In each of these two cases I considered why the weaponisation of doubt in this way was so harmful to us as individuals, as families, as societies and to humanity itself (it sounds a bit grand I know, but climate change is that important).

The beautiful thing about calling a book a 'psychological exploration' is that it signals that the journey is unlikely to be over for some time soon. There's clearly a lot more exploring to do. Doubt is clearly not all bad or all good. It resides in people's lives – sometimes safeguarding them, sometimes punishing them. But when it punishes it can really, really hurt. It can arise from different sources as we have seen, but how it becomes part of our self-identity and a recurrent feature of our life remains to be explored more fully. Sometimes, the origin of the first episode of severe doubt can be localised to a specific situation – a 'difficult' decision about schoolwork, a specific traumatic event (for Picasso), particular dismissive paternal comments (in Kafka's case), a new challenging academic experience (for Pauline Clance), a dream (for Jung), but sometimes it cannot be so clearly localised in this way – like fear of failure passed down through the generations or the creeping impostor syndrome brought on by certain parenting styles. But it can still dominate our lives.

Doubt connects to so much of what we study in psychology – obsessive-compulsive disorder, anxiety, magical thinking, self-efficacy, identity, social interaction, parenting styles, achievement motivation and fear of failure. Its importance can also be seen in those areas where doubt has been deliberately employed as a strategy – that is where doubt has been weaponised to promote smoking and climate change *inaction* – two more significant areas for psychology – addiction and behaviour change. Doubt would seem to be a central concept for so much of psychology but it has been oddly neglected in the past by psychologists, which is especially surprising given its ubiquitous nature and far-reaching influence. However, it has not been neglected by novelists, writers and artists who have used it to give us a glimpse into the minds of their characters or indeed themselves. They have recognised it is a major part of who we are.

But perhaps doubt will not be ignored so easily in the future by those who want to understand the importance of doubt for everyday life and everyday psychology. Doubt is a major part of all our lives in one way or another. Considering it in more detail in this book reminds us, if we need reminding, that the thin, pale, sickly boy from Prague was not alone. He tried to explain how it felt, and he attempted to document its effects and its origin in that most personal of letters – a letter that never reached its intended recipient and could well have been lost for all time. Now we may appreciate what he was telling us and understand the insidious and creeping influence of doubt in his life – and perhaps recognise how it has been operating in our own life as well. With a little bit of reflection and self-analysis, we too may even be able to start to uncover what lies beyond it and where it has been leading us, in order to recognise its full significance and power. We may even be able to identify when it has been used against us and by whom (of course, I'm thinking of the tobacco and energy companies here) and at what personal and societal cost. And that would, *undoubtedly*, be a step in the right direction.

References

Bandura, A. (1977) Self-efficacy: Toward a unifying theory of behavioral change. *Psychological Review*, 84, 191–215.

Bandura, A. (1986a) *Social Foundations of Thought and Action: A Social Cognitive Theory*. Englewood Cliffs, NJ: Prentice-Hall.

Bandura, A. (1986b) The explanatory and predictive scope of self-efficacy theory. *Journal of Clinical and Social Psychology*, 4, 359–373.

Bandura, A. (1990) Reflections on non-ability determinants of competence. In R.J. Sternberg and J. Kolligian, Jr. (Eds.), *Competence Considered* (pp. 315–362). New Haven, CT: Yale University Press.

Bandura, A. (1997) *Self-Efficacy: The Exercise of Control*. New York: Freeman.

Bandura, A., Reese, L. and Adams, N.E. (1982) Microanalysis of action and fear arousal as a function of differential levels of perceived self-efficacy. *Journal of Personality and Social Psychology*, 43, 5–21.

Bateson, G. (1973) *Steps to an Ecology of Mind: Collected Essays in Anthropology, Psychiatry, Evolution, and Epistemology*. New York: Paladin Books.

Bateson, G. (2000) *Steps to an Ecology of Mind*. New York: Ballantine.

Beattie, G. (1983) *Talk: An Analysis of Speech and Non-verbal Behaviour in Conversation*. Milton Keynes: Open University Press.

Beattie, G. (1986) *Survivors of Steel City*. London: Chatto & Windus.

Beattie, G. (1992) *We Are the People: Journeys Through the Heart of Protestant Ulster*. London: Heinemann.

Beattie, G. (1996) *On the Ropes: Boxing as a Way of Life*. London: Victor Gollancz.

Beattie, G. (1998a) *Hard Lines: Voices From Deep Within a Recession*. Manchester: Manchester University Press.

Beattie, G. (1998b) *Head to Head: Uncovering the Psychology of Sporting Success*. London: Victor Gollancz.

Beattie, G. (2002) *The Shadows of Boxing: Prince Naseem and Those He Left Behind*. London: Orion.

Beattie, G. (2003) *Visible Thought: The New Psychology of Body Language*. London: Routledge.

Beattie, G. (2004) *Protestant Boy*. London: Granta.

Beattie, G. (2010) *Why Aren't We Saving the Planet? A Psychologist's Perspective*. London: Routledge.

Beattie, G. (2011) *Get the Edge: How Simple Changes Will Transform Your Life*. London: Headline.

Beattie, G. (2013) *Our Racist Heart? An Exploration of Unconscious Prejudice in Everyday Life*. London: Routledge.

Beattie, G. (2016) *Rethinking Body Language: How Hand Movements Reveal Hidden Thoughts.* London: Routledge.

Beattie, G. (2018a) *The Conflicted Mind: And Why Psychology Has Failed to Deal With It.* London: Routledge.

Beattie, G. (2018b) Optimism bias and climate change. *British Academy Review*, 33, 12–15.

Beattie, G. (2021) *Selfless: A Psychologist's Journey Through Identity and Social Class.* London: Routledge.

Beattie, G. and Bradbury, R.J. (1979) An experimental investigation of the modifiability of the temporal structure of spontaneous speech. *Journal of Psycholinguistic Research*, 8, 225–248.

Beattie, G. and Ellis, A. (2017) *The Psychology of Language and Communication Classic Edition.* London: Routledge.

Beattie, G., Marselle, M., McGuire, L. and Litchfield, D. (2017) Staying over-optimistic about the future: Uncovering attentional biases to climate change messages. *Semiotica*, 218, 22–64.

Beattie, G. and McGuire, L. (2011) Are we too optimistic to bother saving the planet? The relationship between optimism, eye gaze and negative images of climate change. *International Journal of Environmental, Cultural, Economic, and Social Sustainability*, 7, 241–256.

Beattie, G. and McGuire, L. (2015) Harnessing the unconscious mind of the consumer: How implicit attitudes predict pre-conscious visual attention to carbon footprint information on products. *Semiotica*, 204, 253–290.

Beattie, G. and McGuire, L. (2016) Consumption and climate change: Why we say one thing but do another in the face of our greatest threat. *Semiotica*, 213, 493–538.

Beattie, G. and McGuire, L. (2018) *The Psychology of Climate Change.* London: Routledge.

Beattie, G. and McGuire, L. (2020) The modifiability of implicit attitudes to carbon footprint and its implications for carbon choice. *Environment and Behavior*, 52, 467–494.

Boulter, C., Freeston, M., South, M. and Rodgers, J. (2014) Intolerance of uncertainty as a framework for understanding anxiety in children and adolescents with autism spectrum disorder. *Journal of Autism and Developmental Disorders*, 44, 1391–1402.

Boyd, H.W. and Levy, S.J. (1963) Cigarette smoking and the public interest opportunity for business leadership. *Business Horizons*, 6, 37–44.

Bravata, D.M., et al. (2019) Prevalence, predictors and treatment of impostor syndrome: A systematic review. *Journal of General Internal Medicine*, 35, 1252–1275.

Brown, R. and Kulik, J. (1977) Flashbulb memories. *Cognition*, 5, 73–99.

Bullock, A. (2004) *The Secret Sales Pitch: An Overview of Subliminal Advertising.* San Jose, CA: Norwich.

Burney, L.E. (1959) Smoking and lung cancer: A statement of the Public Health Service. *Journal of the American Medical Association*, 171, 1829–1837.

Cheskin, L. (1951) *Color for Profit.* New York: Liverlight.

Clance, P.R. and Imes, S.A. (1978) The impostor phenomenon in high achieving women: Dynamics and therapeutic interventions. *Psychotherapy: Theory, Research and Practice*, 15, 241–247.

Clance, P.R. and O'Toole, M.A. (1987) The impostor phenomenon: An internal barrier to empowerment and achievement. *Women in Therapy*, 6, 51–64.

Clay, C. (1964) *From Harvey Jones and Jesse Bowdry Appearance on CBS 'I've Got a Secret'*, February 24, 1964.

Collins, J.I. (1982) Self-efficacy and ability in achievement behavior. *Paper presented at the Annual Meeting of the American Educational Research Association.* New York.

Conway, M. (1994) *Flashbulb Memories*. London: Psychology Press.

Corbin, C.B. (1972) Mental practice. In W.P. Morgan (Ed.), *Ergogenic Aids and Muscular Performance*. New York: Academic Press.

Correia, M.E. and Rosado, A. (2018) Fear of failure and anxiety in sport. *Analise Psicologica*, 1, 75–86.

Crane, F.G. and Crane, E.C. (2007) Dispositional optimism and entrepreneurial success. *Psychologist-Manager Journal*, 10, 13–25.

Dichter, E. (1947) *The Psychology of Everyday Living*. New York: Barnes & Noble.

Dichter, E. (1960) *The Strategy of Desire*. London: Transaction.

Dunn, G. and Dunn, J. (2001) Relationships among the sport competition anxiety test, the sport anxiety scale, and the collegiate hockey worry scale. *Journal of Applied Sport Psychology*, 13, 411–429.

Edwards, P.W., Zeichner, A., Lawler, N. and Kowalski, R. (1987) A validation study of the Harvey Impostor Phenomenon Scale. *Psychotherapy*, 24, 256–259.

Ehrenreich, B. (2010) *Smile or Die: How Positive Thinking Fooled America and the World*. London: Granta.

Elliot, A.J. and Thrash, T.M. (2004) The intergenerational transmission of fear of failure. *Personality and Social Psychology Bulletin*, 30, 957–971.

Eysenck, H.J. (1966) *Smoking, Health and Personality*. London: Four Square.

Feltz, D.L. and Landers, D.M. (1983) The effects of mental practice on motor skill learning and performance: A meta-analysis. *Journal of Sport Psychology*, 5, 25–27.

Festinger, L. (1957) *A Theory of Cognitive Dissonance*. Evanston, IL: Row, Peterson.

Festinger, L., Riecken, H.W. and Schachter, S. (1956) *When Prophecy Fails*. New York: Harper & Row.

Freud, S. (1900/1997) *The Interpretation of Dreams*. London: Wordsworth Editions.

Freud, S. (1909/1996) *Three Case Histories*. New York: Simon & Schuster.

Freud, S. and Andreas-Salome, L. (1966/1972) *Sigmund Freud and Lou Andreas-Salome Letters*. New York: Harcourt Brace.

Gadsby, S. (2021) Impostor syndrome and self-deception. *Australasian Journal of Philosophy*, 100, 247–261.

Gifford, R. (2011) The dragons of inaction: Psychological barriers that limit climate change mitigation. *American Psychologist*, 66, 290–302.

Gifford, R., et al. (2009) Temporal pessimism and spatial optimism in environmental assessments: An 18-nation survey. *Journal of Environmental Psychology*, 29, 1–12.

Global Risk Report. (2016). 11th ed. www3.weforum.org/docs/Media/TheGlobalRisks Report2016.pdf. Accessed 10th January 2018.

Greenfield, N. and Teevan, R. (1986) Fear of failure in families without fathers. *Psychological Reports*, 59, 571–574.

Hodges, A. (2014) *Alan Turing: The Enigma*. London: Vintage.

Holmes, S.W., Kertay, L., Adamson, L.B., Holland, C.L. and Clance, P.R. (1993) Measuring the impostor phenomenon: A comparison of Clance's IP scale and Harvey's I-P scale. *Journal of Personality Assessment*, 60, 48–59.

Independent. (1996) Eysenck took pounds 800,000 tobacco funds. www.independent.co.uk/news/eysenck-took-pounds-800000-tobacco-funds-1361007.html. Accessed 27th February 2017.

Intergovernmental Panel on Climate Change. (1996) *Climate Change 1995: The Science of Climate Change*. J.T. Houghton, L.G. Meira Filho, B.A. Callander, N. Harris, A. Kattenberg and K. Maskell (Eds.). Cambridge: Cambridge University Press.

Intergovernmental Panel on Climate Change. (2007) *Climate Change 2007: The Physical Science Basis*. S. Solomon, D. Qin, M. Manning, Z. Chen, M. Marquis, K.B. Averyt, N. Tignor and H.L. Miller (Eds.). Cambridge: Cambridge University Press.

Intergovernmental Panel on Climate Change. (2013) *Climate Change 2013: The Physical Science Basis. Contribution of Working Group I to the Fifth Assessment Report of the Intergovernmental Panel on Climate Change.* T.F. Stocker, D. Qin, G.K. Plattner, M.B. Tignor, S.K. Allen, J. Boschung, A. Nauels, Y. Xia, V. Bex, P.M. Midgley (Eds.). Cambridge: Cambridge University Press.

Intergovernmental Panel on Climate Change. (2015) *Climate Change 2014: Mitigation of Climate Change. Working Group III Contribution to the Fifth Assessment Report of the Intergovernmental Panel on Climate Change.* O. Edenhofer, R. Pichs-Madruga, Y. Sokona, J.C. Minx, E. Farahani, S. Kadner, K. Seyboth, A. Adler, I. Baum, S. Brunner, P. Eickemeir, B. Kriemamm, J. Savolainen, S. Schlömer, C. von Stechow and T. Zwickel (Eds.). Cambridge: Cambridge University Press.

Intergovernmental Panel on Climate Change. (2015) *Climate Change 2014: Synthesis Report. Contribution of Working Groups I, II and III to the Fifth Assessment Report of the Intergovernmental Panel on Climate Change.* The Core Writing Team, R.K. Pachauri and L. Maeyer (Eds.). Cambridge: Cambridge University Press.

Isaacowitz, D.M. (2005) The gaze of the optimist. *Personality and Social Psychology Bulletin*, 31, 407–415.

Isaacowitz, D.M. (2006) Motivated gaze: The view from the gazer. *Current Directions in Psychological Science*, 15, 68–72.

Jung, C.G. (1939/2003a) *The Spirit in Man, Art and Literature*. London: Routledge.

Jung, C.G. (1951/1976) *C.G. Jung Letters*. Princeton, NJ: Princeton University Press.

Jung, C.G. (1953/2003b) *Four Archetypes*. London: Routledge.

Jung, C.G. (1961/1973) *Memories, Dreams, Reflections*. London: Collins.

Kafka, F. (1919/2011) *Letter to Father*. Prague: Vitalis.

Kahneman, D. (2011) *Thinking, Fast and Slow*. London: Penguin.

Kasinger, C. (2018). The mind meld of Bill Gates and Steven Pinker. *New York Times*. www.nytimes.com/2018/01/27/business/mind-meld-bill-gates-steven-pinker.html. Accessed 16th March 2018.

Kazdin, A.E. (1978) Covert modelling: The therapeutic application of imagined recall. In J.L. Singer and K.S. Pope (Eds.), *The Power of Human Imagination*. London: Springer.

Kempton, M. (1964) I whipped him and I'm still pretty. *New Republic*. https://newrepublic.com/article/133979/i-whipped-im-still-pretty. Accessed 11th March 2017.

Kolligian, J. and Sternberg, R.J. (1991) Perceived fraudulence in young adults: Is there an 'impostor syndrome'? *Journal of Personality Assessment*, 56, 308–326.

Laporte, G. (1975) *Sunshine at Midnight: Memories of Picasso and Cocteau*. London: Weidenfeld & Nicolson.

Lee, V. and Beattie, G. (1998) The rhetorical organization of verbal and nonverbal behavior in emotion talk. *Semiotica*, 120, 39–92.

Lee, V. and Beattie, G. (2000) Why talking about negative emotional experiences is good for your health: A micro-analytic perspective. *Semiotica*, 130, 1–81.

Lohbeck, A., Grube, D. and Moschner, B. (2017) Academic self-concept and causal attributions for success and failure amongst elementary school children. *International Journal of Early Years Education*, 25, 190–203.

Mahony, P.J. (1986) *Freud and the Rat Man*. New Haven, CT: Yale University Press.

Mailer, N. (1997) *Picasso: Portrait of Picasso as a Young Man*. London: Abacus.

Malraux, A. (1974) *La Tete d'Obsidienne*. Paris: Gallimard.

McCright, A.M. and Dunlap, R.E. (2011) The politicization of climate change and polarization in the American public's views of global warming, 2001–2010. *Sociological Quarterly*, 52, 155–194.

McGregor, H.A. and Elliot, A.J. (2005) The shame of failure: Examining the link between fear of failure and shame. *Personality and Social Psychology Bulletin*, 31, 218–231.

Mosing, M.A., Zietsch, B.P., Shekar, S.N., Wright, M.J. and Martin, N.G. (2009) Genetic and environmental influences on optimism and its relationship to mental and self-rated health: A study of aging twins. *Behavioral Genetics*, 39, 597–604.

Nicholls, J.G. (1975) Causal attributions and other achievement-related cognitions: Effects of task outcome, attainment value, and sex. *Journal of Personality and Social Psychology*, 31, 379–389.

Nicholls, J.G. (1984) Achievement motivation: Conceptions of ability, subjective experience, task choice and performance. *Psychological Review*, 91, 328–346.

Oates, J.C. (1987) *On Boxing*. London: Bloomsbury.

Oreskes, N. and Conway, E. (2010) *Merchants of Doubt: How a Handful of Scientists Obscured the Truth on Issues From Tobacco Smoke to Global Warming*. New York: Bloomsbury Press.

Packard, V. (1957) *The Hidden Persuaders*. London: Macmillan.

Pahl, S., Harris, P., Todd, H. and Rutter, D. (2005) Comparative optimism for environmental risks. *Journal of Environmental Psychology*, 25, 1–11.

Patterson, F. and Gross, M. (1962) *Victory Over Myself*. New York: Bernard Geis Associates.

Pennebaker, J.W. (1989) Confession, inhibition and disease. *Advances in Experimental Social Psychology*, 22, 211–244.

Pennebaker, J.W. (1993) Putting stress into words: Health, linguistic, and therapeutic implications. *Behaviour Research and Therapy*, 31, 539–548.

Pennebaker, J.W. (1995) *Emotion, Disclosure and Health*. Washington, DC: American Psychological Association.

Pennebaker, J.W. (1997) Writing about emotional experiences as a therapeutic process. *Psychological Science*, 8, 162–166.

Pennebaker, J.W. (2000) Telling stories: The health benefits of narrative. *Literature and Medicine*, 19, 3–18.

Pettigrew, M.P. and Lee, K. (2011) The 'father of stress' meets 'big tobacco' Hans Selye and the tobacco industry. *American Journal of Public Health*, 101, 411–418.

Pitt, N. (1998) *The Paddy and the Prince: Making of Naseem Hamed*. New York: Four Walls Eight Windows.

Potter, J. (1996) *Representing Reality: Discourse, Rhetoric and Social Construction*. London: Sage.

Rozin, P., Millman, L. and Nemeroff, C. (1986) Operation of the laws of sympathetic magic in disgust and other domains. *Journal of Personality and Social Psychology*, 50, 703–712.

Rozin, P. and Nemeroff, C. (2002) Sympathetic magical thinking: The contagion and similarity heuristics. In T. Gilovich, D. Griffin and D. Kahneman (Eds.), *Heuristics and Biases: The Psychology of Intuitive Judgment*. Cambridge: Cambridge University Press.

Scheier, M.F. and Carver, C.S. (1985) Optimism, coping and health assessment and implications of generalized outcome expectancies. *Health Psychology*, 4, 219–247.

Scheier, M.F., Matthews, K., Owens, J., Magovern, G., Lefebvre, R., Abbott, A. and Carver, C. (1989) Dispositional optimism and recovery from coronary artery bypass surgery: The beneficial effects on physical and psychological well-being. *Journal of Personality and Social Psychology*, 57, 1024–1040.

Schulz, R., Bookwala, J., Knapp, J.E., Scheier, M. and Williamson, G. (1996) Pessimism, age and cancer mortality. *Psychology and Aging*, 11, 304–309.

Segerstrom, S.C., Taylor, S.E., Kemeny, M.E. and Fahey, J.L. (1998) Optimism is associated with mood, coping, and immune change in response to stress. *Journal of Personality and Social Psychology*, 74, 1646–1655.

Seligman, M. (2002) *Authentic Happiness: Using the New Positive Psychology to Realize Your Potential for Lasting Fulfilment*. New York: Atria.

Selye, H. (1976) *The Stress of Life*. New York: McGraw-Hill.

Sharot, T. (2012) *The Optimism Bias: Why We're Wired to Look on the Bright Side*. London: Constable & Robinson.

Shelley, P.B. (1826) *"Ozymandias". Miscellaneous and Posthumous Poems of Percy Bysshe Shelley*. London: W. Benbow.

Singh, S. (1993) Hostile press measure of fear of failure and its relation to child-rearing attitudes and behavior problems. *Journal of Social Psychology*, 132, 397–399.

Smith, R.E., Smoll, F.L. and Schutz, R.W. (1990) Measurement and correlates of sport-specific cognitive and somatic trait anxiety: The sport anxiety scale. *Anxiety Research*, 2, 263–280.

Star. (2015) Brendan Ingle rejects reunion approach from Naseem Hamed. www.thestar.co.uk/news/no-thanks-naz-brendan-ingle-rejects-reunion-approach-from-naseem-hamed-1-7310471. Accessed 3rd March 2017.

Stipek, D.J. and Gralinski, J.H. (1991) Gender differences in children's achievement-related beliefs and emotional responses to success and failure in mathematics. *Journal of Educational Psychology*, 83, 361–371.

Suskind, R. (2004) Without a doubt. *New York Times*, 17th October 2004.

Turing, A. (1936) On computable numbers, with an application to the Entscheidungsproblem. *Proceedings of the London Mathematical Society*, 43, 230–265.

Turing, A. (1950) Computing machinery and intelligence. *Mind*, 236, 433–460.

Turing, D. (2016) *Prof: Alan Turing Decoded*. Stroud: History Press.

UK climate change risk assessment 2017 synthesis report: Priorities for the next five years (2016). www.theccc.org.uk/tackling-climate-change/preparing-for-climate-change/climate-change-risk-assessment-2017/. Accessed 1st August 2016.

Voyles, M. and Williams, A. (2004) Gender differences in attributions and behaviour in a technology classroom. *Journal of Computers in Mathematics and Science Teaching*, 23, 233–256.

Want, J. and Kleitman, S. (2006) Impostor phenomenon and self-handicapping: Links with parenting styles and self-confidence. *Personality and Individual Differences*, 40, 961–971.

Wilson, T.D., Damiani, M. and Shelton, N. (2002) Improving the academic performance of college students with brief attributional interventions. In J. Aronson (Ed.), *Improving Academic Achievement*. New York: Academic Press.

Woodward, K. (2004) Rumbles in the jungle: Boxing, racialization and the performance of masculinity. *Leisure Studies*, 23, 5–17.

Wooffitt, R. (1992) *Telling Tales of the Unexpected: The Organisation of Factual Discourse*. Hemel Hempstead: Harvester Wheatsheaf.

World Health Organization (2017) *Climate Change and Human Health*. www.who.int/globalchange/en/. Accessed 15th March 2018.

World Resource Institute. (2014). www.wri.org/blog/2014/05/history-carbon-dioxide-emissions. Accessed 10th January 2018.

Index

77 Sunset Strip 106–107

academic work, handling 52
achievement: events, threats 104; focus 63; self-imposed standards 43
affective self-efficacy 82
agency, rejection 150
aliens: curiosity 12; experience, sharing (avoidance) 11; meeting 16, 151; return 10–11; secrets, telling 12; sighting 10
Ali, Muhammed 9, 86, 99
Andres-Salome, Lou 1
anonymity, impact 51
anthropological science, importance (increase) 115
anticipatory avoidance 148
anxiety 2, 151; allaying 37; coping, problem 133; reduction 145; separation anxiety 121
apprehension 2
archaic psychic components 37–38
arrogance, self-serving attributional platform (basis) 102
artificial intelligence (AI) 77
associative connection 118–119
asthma, suffering 109
atmospheric constituent concentration (modification), human activities (impact) 137
atmospheric pollution, importance 131
attention, processes 55
attitudes, change (impossibility) 96
attributional focus 63
attributional style 46–47, 102
authority: figures, opinions (acceptance) 151; symbol 48
avoidance-based goals 88, 104

background knowledge, requirement 51–52
bad events, blame 63

bad news: anticipatory avoidance 148; remembering 62
Bailey, Stu 106
Barrera, Marco Antonio 101
behaviour: control 90; reason, understanding 116
Bernard, Claude 129
body: function 24; mind, influence 125–126
Bogart, Humphrey 106
boxing: culture 152; doubt, impact 104
boxing fights: losses 102; problems 101–102; readiness 100–101
brain: activity (measurement), fMRI (usage) 144; function 24; images, interest 25; machine brain, design 81
brainwashing 103
Brassai, Picasso (interaction) 58, 66–67
brightness, parent definition (internalization) 45
bronchitis, suffering 109
Brown, Roger 62
bullying, allowing 104
Burney, Leroy E. 111
Bush, George W. 10
businesses, closure 139

Camay ad, construction 118–119
cancer, fear (cure) 11
capabilities, faith (loss) 82
carbon dioxide emissions, reduction 136
car buying, women role (Dichter viewpoint) 120
Cartesian doubt 1
Casablanca (movie) 106
Casagemas, Carles (suicide) 69
Castle, The (Kafka) 3
cause-and-effect relationship 59
certainty: absence 139–140; oddness 12–13
change: attributional focus 63; preaching 21
cheats, feeling 6

Cheskin, Louis 113–114
children: achievements, father (narcissistic involvement) 52, 55; impostor phenomenon levels 52; performance-avoidance goals, positive prediction 88–89
Christianity, Jung rejection 39, 150
Chrysler, ads 119–120, 124–125
Chuck, Charlie 108
cigarettes: coffin nails, naming 124, 133; lighter, discovery 123–124; lighting, power 123; roles/functions, range (understanding) 122; smoking, stimulation 128
city folks, psychology 34
Civil Rights Movement: joining 22; protest 23
Clance, Pauline 40–41, 43–45, 53–55, 149; family dynamic assumptions 46; men, impostor fears 47; Oberlin College journey 48
Clay, Cassius 68, 99–100
climate change: arguments 141; counter-narratives 147–148; denial, scientific illiteracy (impact) 146; doubt, relationship 139–140; doubt, weaponisation 148; effects, mitigation 148; effects, witnessing 138; emergency, doubt (impact) 148; human beings, contribution 136; large-scale behavioural adaptation 138; optimist/non-optimist, eye gaze fixations (hotspot analysis) *143*; optimist/non-optimist, scan paths *143*; political/economic implications, media attention 146; problem 148; sceptic arguments, presentation 140–141; science, belief (divisions) 146
Climategate scandal 138
climate system, warming (IPCC conclusion) 137–138
Coggon, David 83
cognitive activity, indicators 48–49
cognitive bias 63–64; optimism bias, association 144
cognitive dissonance: handling 14; impact 16; role 13
collateral damage 18–19
Color Research Institute of America 113–114
comeuppance 92
complete life, doubt/insecurity (importance) 15
Compton Advertising Agency, soap (promotion) 117
Comrie, Bernard 49
concentration, testing 22
confidence 90; absence 6; building 94, 96–97; requirement 52; undermining 129
conflict of interest 133

confrontations, observation 92
Connolly, James 109–110
consciousness 64
consequentiality, impact 63
consumer behaviour, irrational/unconscious side (uncovering) 114
control, processes 55
Council for Tobacco Research (CTR), creation 126–127
Criminal Law Amendment Act (1885) 84
Cubism, range 69

Damocles' sword 40
danger, signal 17
decision making: consequences 27; inhibition 2
defeat: description 102–103; observation 102
depression, commonness 69–70
depressive symptomatology 53
depth approach 113
depth interviewing, usage 116
depth interviews, usage 120
Descartes, René 9; Cartesian doubt 1
despair, inducing 133
Dichter, Ernest 111–125, 133; pleasure, understanding 135; psychoanalytic methods, application 134; psychological research technique, proposal 119; smoking, psychology 125–126
direct empirical knowledge, importance 9
disconfirmation, proselytising (increase) 13
disease, opinion (differences) 128
disgrace, continuation 4
dismissive cynicism 22
dispositional optimism: assessment, LOT-R (usage) 141–142; optimism bias, relationship 142, 144
dispositional optimists, fixation duration 142
dissonance: presence 13; reduction 13
doomsday cult, ethnographic research 13
double bind 4
doubt: absence 19, 57; affliction 8; aloneness 17; APA definition 6; appearance 89–90; attention 153; blame 150–151; climate change, relationship 139–140; components 6–7; connections 152–153; coping 74; defining 1, 15; discussion, rarity 14–15; display 8; doubt-free life, perils 75; doubt-ridden mental life, awareness 63; driver 2, 35, 149–150; emotional/cognitive components 28; forced removal 81; forms 8; idiographic approach 149; immunity 28; impact 15–16, 34–35; importance 2, 15, 28, 30, 104; inhibitor,

impact 85; initiation 3; internalisation 23; manufacturing 106; "merchants of doubt" 135; nature/meaning 1; origin 8, 27; personal aspect 150; precluding 12; presence 8–9; private type 40; rational thought, instrument 2; refuge 145; removale 84; rules, impact 77; scientific doubt, presence 132; secret, impact 38; self-doubt, appearance 82; spread 5, 28; therapeutic interventions 9–10; treatment 86; understanding 74; unpleasantness 31–32; weaponisation 147, 152
doubting: habits, change 9; variation 16
Doubting Thomas 9
dreams: image, strangeness 38; investigation 34
Duff, Mickey 103
dying, probability (estimations) 61
Dylan, Bob (songs) 21
dynamic principle 115–116
dyslexia/dyscalculia 58
dysphoric personality tendencies 53–54

early family history, expressions 45
efficacy: bias causal attributions, self-beliefs 82; self-efficacy 2, 7, 82, 85, 148; sense, resilience 81
ego: control, reduction 114; splitting 121
ego-personality, representation 30
Electronic Brain, Turing work 83–84
Eleven Plus 22–23
embarrassment 88; suffering 104
emotion: expression 5–6; handling 117; involvement 61
emotional experiences, articulation/sharing 5–6
emotional self-efficacy 82
energy demands, patterns (change) 136
entrepreneurial success, psychological factor 144
ethics, doubt (importance) 15
Eubank, Chris 93
events: handling 115; judgement 61
excellence, pursuit (Turing) 80–81
existence, confirmation 5
experience, role 36
experiences, impact 61–62
extraverts, introverts (differences) 128
eye-gaze fixations: eye tracking, usage 141; hotspot analysis 142, *143*
Eysenck, Hans 126, 127–135

failures: attributions 47; characterisation 102; contemplation 60; external attributions 82; flashbulb memories 63–64; internalisation, avoidance 74; remembering 62
failures, fear 2, 88, 104; blocking 105; evidence 88–89
families: decision-making role 120; failure, fear (passing) 105; family myth, belief 44
family dynamics 45–46; assumptions (Clance) 46; identification 47
fantasies, projection 120
father: death, impact 26–27; death, remembering 62–63; murder, child desire 67; narcissistic involvement 52–53; overprotection, paternal care absence (correlation) 52; rejection (Picasso) 67; repulsive tyrant 150; role, Kafka theory 56; role, significance 52; unconsciousness/coma 24–25
faults, accusation 60–61
Fawkes, Guy 95
fear: fear-inducing message, impact 147; language, usage 102; presence 89; processes 55; visibility 97
feedback: effort 81; monitoring 136
feelings: guilt feelings, assuaging 125–126; statements, discrepancy (detection) 116
feminine charms, impact 44
Festinger, Leon 13–14
First Communion (Picasso) 66
flashbulb memories: emotional reaction 62; failure, relationship 63–64; formation, prevention 63–64; immunity, absence 63; importance 62; research, basis 64
Franklin, Benjamin 137
fraud: committing 40; feeling 6, 40, 54; perception 53–54; self-perceptions 54
Freud, Sigmund: academic world, relationship 31; attack (Jung) 31; daemon possession 35; doubts (Jung) 30; friendship (Jung) 35; obsessional neurotics, uncertainty (need) 1; philosophical premises, absence 33; philosophy, knowledge (absence) 33–34; psychoanalysis (Jung) 32; repression mechanism, articulation (Jung agreement) 31; reputation, damage 31; sexual experiences, symptoms (derivation) 1–2; sexual obsession 38; sexual theory, emotional involvement 32
fright 100–101; impact 62
functional magnetic resonance imaging (fMRI), usage 144
functional principle 115; application 117
fundamental insights, principle 116
future, stress/anxiety (reduction) 145

Gates, Bill 146
Get the Edge (Beattie) 63
global warming: changes 136; occurrence, belief 138
glucose, reserves (release) 60
Graham, Herol 91, 93, 94
greenhouse gas emissions: changes 136; difference, making 148; increase 137
guilt: feelings 119; feelings, assuaging 125–126; ridding 117–118; washing 119
guilt-free caress, importance 117–118

Hamed, Naseem (Naz) (Motormouth) (the Prince) 9, 87–93, 97, 100–104; comeuppance/failure 92–93; defeat, description 102–103; Ingle, friendship (loss) 103
Harris, Beth 71
health damage, risk (estimates) 143
Hidden Persuaders, the (Packard) 113
Hill, John 131–132; tobacco company meeting 131
historical narrative, impact 32
Hoffman, Andrew 146
homesickness 110
homogeneous entity, doubt 39
How Culture Shapes the Climate Change Debate (Hoffman) 146
human actions, motivations 116
human activity, changes 136
human beings, rational creatures 112
human motivations: change 115–116; research 116; understanding/measuring, Dichter approach 115
human weakness, arguments 110–111
Hussein, Saddam 10
hypochondria 5
hysterical symptoms, affective/unconscious ideas (impact) 34–35

illnesses (increase), climate change (impact) 138
imagistic memories, impact 73
Imes, Suzanne 40, 41, 44–45, 55; family dynamics, assumptions 46; men, impostor fears 47
imitation game 79
impostor feelings 6, 51; derivation, self-criticism (impact) 56; development, father role (importance) 52–53; self-handicapping tendencies, correlation 52

impostor phenomenon 54; expression, usage 53; features 42–43; feelings 44; incidence 43–44; rediscovery 52–53; studies, review 48; suffering, observations 45
Impostor Phenomenon Scale (IPS), construct validity (criticism) 53
impostors: grouping 44; self-consideration 41; self-declared impostors, fear 42
impostor syndrome 2, 3; existence 53; Kafka suffering 54; presence 55; self-destruction, possibility 55; self-handicapping tendencies, correlation 55
impression-management 53
impression management skills, self-monitoring skills (combination) 53–54
indecision, perception 2
industrial revolution, impact 136
Ingle, Brendan 88–104, 152; bullying, allowing 104; failure, fear (blocking) 105; Hamed, friendship (loss) 103
Ingle, John 97
injurious consequences, cause 126
innate ideas, Platonic concept 34
inner critic, development 52
inner life, documentation 30
innermost thoughts, representation 29
inner voice, impact 7–8
insecurity, importance 2, 30
intellectual impostors 42
intelligence, meaning (understanding) 43
intercesseurs 71–72
Intergovernmental Panel on Climate Change (IPCC), reports/consensus statements 137
Interpretation of Dreams, The (Freud) 30, 34
introverts, extraverts (differences) 128
investigatory procedure, usage 27
Iraq, invasion 10
Irish, English opinion 95
Irish Republic Army (IRA), impact 22
isolation (feelings), cigarettes (usage) 122, 134
I, the King (Picasso) 8, 57
I've Got a Secret (Ali) 99
Ivory soap, promotion 117–119

Josephson, Brian 49–51
J. Stirling Getchell, Inc. 119–121
Jung, Carl 149; archetypes, concept/theory 34, 39; child development 38; childhood dream 36; Christianity, rejection 39; collective unconscious, theory 39; doubt/insecurity, life components 15;

doubts 30; doubts, consideration 31; ego-personality, representation 30; father, religious collapse 38; Freud friendship 35; inner life, documentation 30; innermost thoughts, representation 29; intellectual life, unconscious beginnings 36; oppression 37; passive-aggressive style 33; personal attacks, usage 33; personal/ psychological development, truth 35; rebirth 30; ritual phallus, understanding 35–36; second personality 31–32; self-narrative, clarity 38; sexual symbolism 38; unconscious rituals, engagement 38
Jung, Carl (dream) 29; description 35; secret 36–37, 39; secret, physical manifestation 37

kabir (manikin) 37–38
Kafka, Franz 149; Damocles' sword 40; disgrace 4; double bind 4; doubt, origin 3; doubt, spread 5; failure, fear 41; fear, educational background (basis) 41; hypochondria 5; impostor syndrome, suffering 54; inner voice, impact 7–8; neuroses 5; self-doubt 15–16; self-doubt, origin 3–4, 47–48
Kafka, Hermann (impact) 3–4; child-rearing devices 4
Kahneman, Daniel 67–68, 73
kidology 97, 152
King of the World (Remnick) 97
Kulik, James 62
kuntu, belief 70–71
Kyoto Protocol 146

Ladd, Alan 121
land use, patterns (change) 136
language, interest 48–49
Lanzer, Ernst 1
Laporte, Geneviève 68
law, doubt (importance) 15
lecturing, joy 62–63
Les Demoiselles d'Avignon (Picasso) 65, 71, 73
life: change, absence 91–92; doubt-free life, perils 75; doubt/insecurity, importance 30; flow (disruption), doubt (impact) 28; loss 104; smoking, impact 111
"Linen Slaves of Belfast" (Connolly) 109–110
Liston, Sonny 97, 99–100
logogen model 49
loneliness (feelings), cigarettes (usage) 122, 134

longevity, increase 145
LOT-R (psychometric scale), usage 141–142
lungs: cancer risk (relationship), statistical data (problem) 130; cancer/smoking, relationship (epidemiological evidence) 129–130; mechanical irritation 109

machine brain, design 81
magical thinking 150; laws 72
Mailer, Norman 99
Malraux, André 71
man-made climate change, con (belief) 140
marketing, depth approach 113
marriage, convertibles (threat) 120
masks, Picasso appreciation 71
Mead, Margaret 115
Memories, Dreams, Reflections (Jung) 35
memory: fragments, return 37; trace 36
men, impostor fears 47
Merchants of Doubt (Oreskes/Conway) 131
"merchants of doubt" 135
message, mind (rejection) 111
Metamorphosis, The (Kafka) 3
Mills, Mick 91, 97
mind (settling), running (impact) 76
Morcom, Christopher 79–81, 85
morphogenetics, Turing interest 79
Morton, John 49
mother, pain 25–26
Murray, Arnold 83–84
Murray (nee Clarke), Joan (Turing marriage proposal) 77–78
Museum of Modern Art (MoMA), artworks (critique) 64–65

narrative: construction 12; importance 57–58
National Front 89, 94
negative evaluation 51–52
negative experiences, connection (absence) 27
negative feeling, anticipation 6
negative life events: experiencing 144; underestimation, impact 145
negative maternal characteristics 88–89
negative moments/experiences, remembering 62
Nelson, Johnny 93, 94, 97
Nemeroff, Carol 69
nervousness, physiological feelings (enjoyment) 63
neuroses: cure 134; formation, sex/ repression (role) 36

neuroses (Kafka) 5
non-optimists: climate change, impact 144; eye gaze fixations *143*; fixation length 142; scan paths *143*
nursery rhymes, public recitation 90–91

Oates, Joyce Carol 89
obsessional neurotics, uncertainty (need) 1
obsessive-compulsive disorder 1
Oedipus complex 67–68; Freud concept 68
"On Computable Numbers" (Turing) 76
optimism bias: cognitive biases, association 144; dispositional optimism, relationship 142, 144
optimists: climate change, impact 144; eye gaze fixations, hotspot analysis *143*; scan paths *143*
oral gratification 123
Orange Order 19; joining 22, 23
orders, obeying/disobeying 4
O'Toole, Slugger 95
overt conscious thought 27
Oxymandias 30

Packard, Vance 113
parents, self-handicapping 104
Parkinson's disease 20–21
paternal care (absence), father overprotection (correlation) 52
Patterson, Floyd 97, 100
Patterson, Mary Jane 41
peek-a-boo, instruction 59–60
personal doubt, origin 7
personality: concept 118; release 121; types, hypothesis (Eysenck) 128
personal memories, importance 64
philosophy: doubt, importance 15; knowledge, absence (impact) 34
phony, discovery 41–42
phthisis, suffering 109
physiological changes 61
Picasso Lopez, Maria 58
Picasso, Pablo 57, 149; artistic gift 58; Blue Period 68–69; Brassaï, interaction 58, 66–67; Cubism, range 69; cultural evolution 72; depression, suffering 68; dust of Paris, deprivation 68; dyslexia/ dyscalculia 58; father, rejection 67; fingernail clippings, saving 69; genius, perception 66; *intercesseurs* 71–72; masks, appreciation 71–72; outsider status 66; productivity, explanation 70; rage 70; reaction formation 70; risk taking/ confidence 73; self-reinvention 67;

self-styling 68; superstitions 74; terror, escape 58–59
Pinker, Stephen 146
Pitt, Nick 103
pleasure: loss 131; oral pleasure, understanding 135
politics, doubt (importance) 15
positive feedback, handling 43
positive feel-good moments, remembering 61–62
positive thinking, power 145
power, symbol 48
predictability, desire 77
prejudice, absence 99
pride: excess 20; loss 100
Priestley, J.B. 39
prime positive memories, usage 63–64
probabilities, scientific language 148
problems (probability), imagistic memories (impact) 73
products: personality/image, concept 118; symbolic importance, understanding 114; thinking/feeling, understanding 134
prophecy, failure (cult reaction) 13
Protestants, singing 21
Protestant Ulster Volunteer Force (UVF): impact 22; re-establishment 18
pseudoscientific facts, assembly 120
psyche: depths 32; nature, understanding 33–34; organic view 39
psychic development, experience (role) 36
psychic lesion/conflict, impact 31
psychoanalysis: applications, focus 113; founding 33; insights 114
psychoanalytic theory 119
psychoanalytic training 113
psychodrama, usage 116–117
psycholinguistic processes, uncovering 48–49
psychological attachment 121
psychological change 91
psychological coaching 101–102
psychological consequences 7
psychological factors, impact 115
psychological forces 112
psychological issues, presentation 43
psychological training 97
public figure, death 62

qualifiers, usage 63

radiant energy, absorption/scattering 137
rational creatures, absence 61
rationality, increase 146

rational thought, doubt (impact) 2, 15
Rat Man, study 1, 8
reaction formation 70
real experience, mobilisation 116
reason, tribalism (impact) 146
rejection 49
relationships, risk taking (importance) 60
Remnick, David 97
Report on Smoking and Health (Royal College of Physicians) 127–128
repression: mechanism 30–31; mechanism, articulation (Jung agreement) 31; non-sexual origin 32; role 36; sexual origin 32
response efficacy 140
risk: aversion 59; contemplation 60; perception 2; rejection 150
risk taking 59; impact 73; importance 60; increase 63, 74
risky decisions: avoidance, reason 62; confrontation 60–61
ritual phallus, understanding 35–36
Robinson, Steve 97, 101
Roman Catholicism, impact 98–99
routine, rehearsal 95
Routledge, Norman 78
Rozin, Paul 69, 72
Rubin, William 71
Ruiz, Jose 58
Ruiz, Pablo 66
Ruiz, Salvador 57

satisfaction, loss 131
schizophrenics, observation 30–31
science, doubt: existence 141; importance 15, 33
scientific doubt, presence 132
scientific illiteracy, impact 146
scientific truth, doubt (impact) 34–35
second personality 31–32
security, study/work (impact) 26–27
seducer, impact 118–119
seductress, caress 118
selective perception, impact 111
self-assurance, recovery 82
self-belief, building 84
self-confidence, absence 43
self-criticism, imposition 52
self-declared impostors, fear 42
self-depreciation 53–54
self-destruction, impostor syndrome (impact) 55
self-doubt 80–81; appearance 82; importance 104; origin 3–4, 7

self-efficacy 2, 140; doubt 148; growth 151–152; high level 85; low level 7; perception 84; recovery, speed 82
self-esteem: low level 7; negative correlation 54
self-fulfilling prophecy 145
self-handicapping strategies 88
self-handicapping tendencies: imposter feelings, correlation 52; impostor syndrome, correlation 55
self-help book, message 64
self-identity 13
self-image: distortion 42; protection 52
self-monitoring skills 53–54; impression management skills, combination 53–54
self-narrative, clarity 38
self-reflection, usage 150–151
self-respect 88
self-serving attributional platform, basis 102
self-worth, rejection 150
Seligman, Martin 144–145
Selye, Hans 126–127, 132
separation anxiety 121
sex, role 36
sexual experiences, symptoms (derivation) 1–2
sexual symbolism 38
sexual theory (Freud), emotional involvement 32
shame 88; suffering 104
Sheffield, alien visitation capital 13
Shipherd, John J. 41
smoke, chemical composition 130
smoking: action, impact 130; activity 122; advertising 121–122; ailments, blame 129; analysis 115; cessation, rationality 130–131; conflict of interest 133; distaste 108; effects 126–127; harm, evidence 125; health effects, deception 131–132, 135; impact 17; oral pleasure 123; pleasures/satisfaction 132; psychology (Dichter) 125–126; public perception 108–109; quantity, question 130; reduction, efforts 125–126; research efforts/legislative measures 131; risk, correlation problem 130; scientific evidence, challenge 131–132; sophistication, absence 107; statistics, caution 130; unconscious association 125
Smoking, Health and Personality (Eysenck) 127
social communication, understanding (difficulties) 77
social connotations 123
social exchange, initiation 107
social skills 44; problems 77
soul, reintegration 39

speaking, hesitancy/stammer 4
Special Account Number 4 131, 135
Spence, Gusty 18, 21–22
Spencer, Jeff 106
spirits, influence (prevention) 71–72
sports centring, doubt/self-doubt (impact) 88
statements/feelings, discrepancy
    (detection) 116
statistical research, limitations 119
Stewart, Philo P. 41
"Strange Life and Death of Dr Turing, The" 77
stress 126–127; reduction 145
stress-timed gesture 61
study, increase 26–27
success: characterisation 102; internal
    attributions 82; women (success),
    internalisation (inability) 40–41, 55
suffering, remembering 100
survival: flashbulb memories, importance
    62; risk taking, importance 60
swear words, usage (avoidance) 94–95
symbolic connotations 123
symbolic meaning 113–114
sympathetic magical thinking 69

talent: identification 92; reward 59
talk, control 90
Taylor, Edwin 69
technical factors, impact 115
tension differential, impact 115
thinking: forms 54; involvement 61
Thunberg, Greta (environmental message)
    140, 147–148
Tobacco Industry Committee for Public
    Information, creation 131–132
training, demand 75
transmission, chain (breaking) 89
transport, patterns (change) 136
trauma 70; connection, absence 27;
    projection 35
Trial, The (Kafka) 32
tribalism, impact 146
Troubles, the 18
true feelings, mobilisation 116
Trump, Donald: climate scepticism 145;
    fake news message 139–140
Turing, Alan 8, 77, 149; abilities, self-
    doubt 79; Bletchley Park work 76–77;
    computer, term (usage) 76; criticism
    81; death 83; doubts 85; excellence,
    pursuit 80–81; experience, mastery
    81–82; genius, documentation 151–152;
    gross indecency, arrest 78; imitation
    game 79; marriage proposal 77–78;
    morphogenetics 79; numbers, beauty 80;

personality, change 85; self-efficacy
    feelings, growth 82; self-efficacy level,
    elevation 85; sex crime, report 78; sexual
    offences charge, guilty plea 78; truth,
    defense 84–85; writing, problems 79–80
Turing, Dermot 77
Turing Machines 76
Turing Text 79
turn-taking, learning 59–60
Tzara, Tristan 67

UFO, sighting 10
uncertainty 2; feeling, association 33;
    impact 6; paralysis 77; rules, impact 77
unconscious: access/manipulation 134;
    manipulation 112; role 67–68
unconscious association 125
unconscious mind, focus 133–134
unconscious rituals, engagement 38
unconscious symbolism 133
University of Cambridge, seminar attendees
    (evaluation) 49

verbal routines, tone/automaticity 99
visual stimuli, ambiguity 133

war, casualty 18–19
weakness: detection 60; signalling 60
Western societies, socio-psychological
    change (dangers) 145
widow, public view 26
Williams, F.C. 83
Wilson, Sammy 140
wisdom, doubt/challenge 38
Wittgenstein, Ludwig 50
women: ability, overestimation 42; ability,
    self-underestimation 43; brightness,
    parent definition (internalization) 45;
    doubt/fear 42–43; evaluation, dread
    42; impostors, self-evaluation 44;
    impostor syndrome, identification
    55; objective evidence, dismissal 55;
    performance, appreciation 44; phony,
    discovery 42; positive feedback,
    handling 43; precocity 44–45;
    psychological issues, study 43; quality,
    evaluation 41–42; role, Dichter
    viewpoint 120; self-confidence,
    absence 43; self-image, distortion 42,
    55; success, internalisation (inability)
    40–41, 55
work, increase 26–27
worry 2

Zucker, Steven 71

For Product Safety Concerns and Information please contact our EU
representative GPSR@taylorandfrancis.com
Taylor & Francis Verlag GmbH, Kaufingerstraße 24, 80331 München, Germany

9 781032 252049